W9-BIV-384

Life Counts

Tarawa (Kiribati), an island in the South
Seas, January 1, 2000, on the International
Date Line. Captain Raimund Gross
plants the first tree of the third millennium
for Life Counts. The little coconut palm
may flourish.

Life Counts
Cataloguing Life on Earth

**by Michael Gleich, Dirk Maxeiner,
Michael Miersch, Fabian Nicolay**

In Collaboration with:

UNEP
The United Nations
Environment Programme

IUCN
The World
Conservation Union

WCMC
The World Conservation
Monitoring Centre

Aventis

Translated by Steven Rendall

**Atlantic Monthly Press
New York**

Life Counts *project*

The Life Counts Project supports various scientific initiatives and publications about biodiversity, including this volume and the studies on which it is based. This book urges people everywhere to maintain the world's multiplicity of species. Much of the data and figures used were provided by the World Conservation Monitoring Centre (WCMC), which is also publishing simultaneously – and with financial help from the Life Counts Project – "Global Biodiversity", a companion work addressed chiefly to specialists.

All of Life Counts's activities are supported by Aventis (a merger of Hoechst AG and Rhône-Poulenc SA), a global leader in pharmaceuticals with strong European roots. Aventis believes strongly in the United Nations conventions regarding biological diversity and has committed itself to making "sustainable development" possible. Indeed, its two predecessor companies were part of the World Business Council for Sustainable Development. Aventis's international partners in the Life Counts Project are WCMC, the World Conservation Union (formerly the International Union for the Conservation of Nature, IUCN), and the United Nations Environment Programme (UNEP).

The Life Counts Project seeks to show how important it is that we all be concerned with biodiversity. Although it is most important in developing countries in the Southern Hemisphere, its maintenance also benefits industrialized countries in the north. To help transfer scientific knowledge from north to south, Aventis will make

a substantial number of books available at no cost to schools, universities, and nongovernmental organizations in countries where biodiversity "hot spots" are located. In the framework of the Life Counts project, Aventis cooperates with the following international partners:
– World Conservation Monitoring Centre (WCMC)
– The World Conservation Union (IUCN)
– United Nations Environment Programme (UNEP)

Scientific Advisory Panel
– Dr. Mark Collins, Chief Executive, World Conservation Monitoring Centre (WCMC), Cambridge (UK).
– Carl Haub, Senior Demographer, Population Reference Bureau, Washington, D.C., and Consultant to the German World Population Foundation (Deutsche Stiftung Weltbevölkerung).
– Jeffrey A. McNeely, Chief Scientist, The World Conservation Union (IUCN), Gland (Switzerland).
– Prof. Dr. David Pearce, Centre for Social and Economic Research on the Global Environment (CSERGE), London.
– Prof. Dr. Josef H. Reichholf, Zoologische Staatssammlung, Munich; Member of the Governing Board of the World Wide Fund for Nature (WWF) (Germany).
– Bai-Mass Taal, Senior Program Officer, "Biodiversity," United Nations Environment Programme (UNEP), Nairobi.

The members of the advisory board made themselves available to advise and help in their respective areas. However, they do not necessarily agree with the individual contributions to this volume.

The Network for This Book

Authors
(responsible for conception,
production, and coordination)

Michael Gleich
Dirk Maxeiner
Michael Miersch
Fabian Nicolay
(art direction and graphics)

Scientific Research

**World Conservation
Monitoring Centre (WCMC)**
**Dr. Brian Groombridge
(Director), Rachel Cook, Neil
Cox, Martin D. Jenkins,
Ben Warren, Christoph Zöckler**

Dr. Markus Borner
Tanzania Wildlife Conservation
Monitoring, Seronera (Tanzania)

Wolf Krug
Environmental Economist,
Centre for Social and Economic
Research on the Global
Environment (CSERGE),
London

Dr. Simon Mduma
Tanzania Wildlife Conservation
Monitoring, Arusha (Tanzania)

Prof. Dr. Horst Schminke
University of Oldenburg
and the Deutsche Gesellschaft
für Systematik

Guest Authors

Prof. Dr. Klaus Töpfer
Executive Director,
United Nations Environment
Program (UNEP),
"Sustainable Use: A New
Method for Protecting Species"

Jeffrey A. McNeely
Chief Scientist,
The World Conservation Union
(IUCN), "Threatened and
Threatening: Our Love-Hate
Relationship with Nature and Its
Conservation"

**Dr. Maritta R. von
Bieberstein Koch-Weser**
Director General,
The World Conservation Union
(IUCN), Foreword

Prof. Dr. Josef H. Reichholf
Zoologische Staatssammlung
München, "Progress through
Catastrophes: How Extinctions
Further Evolution"

Life Counts *project*

Collaborators

Dr. Monika Offenberger
Biologist, Munich
(Chapter "Extinctions: Losing
Species Before Their Roles
Are Understood")

Silvia Sanides
Biologist and journalist,
Charlottesville, VA
(Chapter "Preserving Diversity:
The Next Fifty Years")

Thomas Weidenbach
Filmmaker and scientific
journalist, Köln
(Chapter "Fair Dealing: Who
Gets Nature's Dividends")

Dr. Birgitt Salamon
Biologist, Munich
(Documentation)

Bernd Muggenthaler
Political scientist and journalist,
Munich (Documentation)

Mini Misra
Communications designer,
Darmstadt (Graphics)

Gabriele Lorenzer
Photographer, Frankfurt
(Double-page black-and-white
photographs)

Gundhild Eder
Scientific graphic artist for
the Senckenberg Natural History
Museum in Frankfurt
(Illustrations)

Ursula Popiolek-Winkler
Darmstadt (Copyediting)

Astrid Leifheit
Digitales, Frankfurt
(Lithography for the double-page
photographs)

**For valuable information
and encouragement we are
very grateful to:**

Prof. Dr. Wilhelm Barthlott
University of Bonn

Prof. Dr. Angelika Brandt
University of Hamburg

**Deutsche Stiftung
Weltbevölkerung (DSW)**
Hannover

Prof. Dr. Bert Hölldobler
University of Würzburg

Prof. Roland Irslinger
FH Rottenburg

Dr. Stephan Krall
Deutsche Gesellschaft für
Technische Zusammenarbeit,
Eschborn

Dr. Johannes Maurer
German Human Genome
Project, Berlin

Prof. Dr. Paul Müller
University of Saarbrücken

Prof. Dr. Gunther Nogge
Köln Zoo

Günter Rupertus
Kurpfälzische Münzhandlung,
Mannheim

Dr. Klaus Schmitt
Deutsche Gesellschaft für
Technische Zusammenarbeit,
Uganda

Dr. Elke Stumpf
Deutsche Gesellschaft für
Technische Zusammenarbeit,
Eschborn

Christian Weber
Redaktion FOCUS

Maps

Mountain High Maps®
Copyright © 1993
Digital Wisdom Inc.

Table of Contents

I. Numbers and Research

Plates

The Wealth of Nature, the Diversity of Cultures

Foreword by the Authors

"In those days, a decree went out from Caesar Augustus that all the world should be enrolled" (Luke 2:1). Thus, according to Christian tradition, a census stands at the beginning of the present era.

More than 2,000 years later, we reflect that while the twentieth century bore the stamp of physics and chemistry, the driving force of the next century will be biology. Humans are therefore entering into a new alliance with their environment, an alliance in which technology and nature must not compete, but rather cooperate. A decrease in the diversity of living creatures also means a decrease in our chances of survival, and so it is in our best interest to revisit that biblical theme of the census – this time extending it to the whole of the biosphere.

Six billion people live on Earth, but we share the planet with quintillions of other life forms ranging from bacteria to whales. They make up life's infrastructure, and are in effect the underpinnings of human existence. We have sought to provide as much data about them as possible, from as many sources as possible. How many birds and how many trees are there per human being? What treasures and resources are hidden behind these figures? Who knows the enormous economic value produced by insects that pollinate plants?

When we contacted scientists, politicians, conservationists, and business people about our project, we had our first surprise: Their willingness to help us was overwhelming. Leading biologists, economists, and demographers made their stores of knowledge available to us. The World Conservation Monitoring Centre (WCMC), which constantly records the latest plant and animal population data, agreed to help us obtain and evaluate information.

To produce their data, WCMC–affiliated research submarines dive to the bottom of the Mariana Trench; microbiologists in Norway analyze bacteria in the soil and zoologists fly above herds of gnus in the East African Serengeti. It is an adventurous and truly global enterprise, a perfect fit for the Life Counts Project.

Our second surprise came with the evaluation of our data. It is astonishing – and frightening – how little we really know about the state of the biosphere. Experts estimate that in tropical forests there are millions of insects that are completely unknown to us. We still have not investigated most of the life forms in the oceans. Even for large mammals we often have only estimated numbers. Nevertheless, the figures presented in this book sketch a picture of life on our planet that is unique and enlightening, contradicting many clichés.

The members of the Life Counts Project hope that this picture – dynamic, colorful, surprising – will create a bridge between science and the general public, promoting new esteem for the natural world and spurring further research. We chose to present our findings in a book because it is the most inclusive of formats, requiring not even an electrical outlet to use, and can be published in many languages throughout the world. We have tried to transcend socioeconomic and cultural boundaries by showing how every culture can learn from our studies and how we can work together toward a common goal.

Life Counts records a moment in the life of the planet. We hope that fifty years from now it will be seen as a first step toward a successful breakthrough into the era of biodiversity.

Munich, December 1999 Michael Gleich, Dirk Maxeiner, Michael Miersch, Fabian Nicolay

Preface

by Stewart Brand

This is one of those books that's tough to have a bedmate reading. "Listen to this! It says here that..."

... all ants weigh the same as all humans.

... there are three hundred species of bacteria in your mouth, dear.

... five thousand species of plant were introduced to the United States from elsewhere; that's 29 percent of all the plants here.

... parts of Europe have more species now than before humans arrived.

... biodiversity can be saved at one-fourth the cost of destructive subsidies.

Understanding is driven by facts (numbers). Policy is driven by facts (numbers). Biology is still a young science because its data collection is still so rudimentary, given the seething diversity and complexity out there. These days I'm working with an organization that aims to identify and catalog all the species on Earth in the next twenty-five years. Two of the scientists on the project – Edward O. Wilson and David Hillis – claim that when all the species are known, "then biology becomes a predictive science."

Life Counts is a report – a superb report – on a great work in progress, far beyond merely finding and naming all the species. We are just starting to ramp up on finally understanding the dynamics of life on Earth, which is the infrastructure of human life. Earth life is 3.5 billion years old. Modern humans have been around for maybe fifty thousand years (i.e., the last 1/70,000th). If we want to be around for the next fifty thousand, it behooves us to learn how life really works, because we're now affecting the whole system, for good or ill.

A data-intensive science like biology is not driven by hypotheses and models so much as it is by tools and toil – ingenious new devices and the passion to apply them globally. GPS locators, the Internet, and ever-faster DNA sequencers are already revolutionizing field biology. The coming of nanotechnology will open up the teeming world of microbial life – where we'll study most of the biomass, most of the metabolism, most of the evolutionary history, and most of the remaining mystery of life on the planet.

It used to be that taxonomists sequestered their data; that was how they acquired power as experts in their particular twig on the tree of life. The Internet has changed all that. The Human Genome Project demonstrated conclusively that discoveries instantly published online in GenBank could move the science much more rapidly, with no loss of quality in the work, because public visibility allowed quick correction of wrong information. Good new data now acquires its value from being widely shared, not from being hidden.

A major element still missing in biological data collection is duration. Megalife lives in megatime, yet our best datasets still measure in minitime – forty years of animal census data from the Serengetti in Africa (see p. 46) is considered to be exceptionally lengthy. We can infer some long-term trends from fossils, from lake and ocean sediments, from packrat middens, from tree rings, but it's not the same as numbers carefully collected in the present and carefully preserved and correlated over decades and centuries. I hope there's a *Life Counts* and *Global Biodiversity* (the specialist companion book) version 2.0 in 2012 and version 10.0 in 2102. That's when the deepest trendlines and cycles will turn up.

When you have data you're confident in, at the right degree of resolution of detail, then you get to do fun science and realistic policy. The fun is figuring out how life works; the responsibility is figuring out what civilization should do with that knowledge.

For instance, to take a current debate, how large is human-caused species extinction, really? How important is it, really? These are knowable! Then: what activities can we change to improve the situation? The better information we have, the more finely tuned our response can be.

As *Life Counts* demonstrates, we have a lot of information already, a good start, enough for a first cut at intelligent public policy. But the best is yet to come, and this book shows why it is so important to press on, and how fascinating it will be.

Stewart Brand, cofounder, All-Species Inventory (www.all-species.org) and The Long Now Foundation (www.longnow.org)

Part One
Numbers and Research

Through a glorious accident of evolution,
our intelligence, we have become the
protectors of the continuing survival of life
on earth. We have not sought this role,
but we cannot decline it. It may be that we
are not up to it, but we must set about it.
Stephen Jay Gould

Human Favorites

How the Stars of the Environment Are Doing:

Scientists have now distinguished 1.75 million species. Most people know of fewer than 0.01 percent of these. The most prominent are the large mammals and birds whose beauty fascinates many people. Children see tigers, elephants, and eagles in books, films, and zoos. But how do people treat their favorite animals? Many of them are endangered species as a result of excessive hunting, and may become extinct – like the tiger and the rhinoceros, which are used in making traditional Chinese medicines, and the panda and the orangutan, whose habitat is growing steadily smaller. Others live in remote areas sheltered from hunting – like the king penguin, which lives in Antarctica. Still others maintain large populations despite the fact that people hunt and use them – such as the white-tailed deer and the saddle seal. Finally, a few can survive because for centuries governments have protected them from over-exploitation – such as the California sea lion and the koala.

1. Bald eagle

Population: around 110,000 to 150,000. Trend: increasing. The national bird of the United States was once endangered by DDT. Since the use of this chemical has been prohibited, the population of bald eagles has grown.

2. Bison

Population: over 200,000. Trend: increasing. Before the arrival of European settlers, there were about 60 million bison. At the beginning of the twentieth century there were fewer than 100.

3. White-tailed deer

Population: about 19.6 million. Trend: sharply increasing. The model for Disney's *Bambi* is one of the commonest wild hoofed animals.

4. California sea lion

Population: over 145,000. Trend: sharply increasing. After the prohibition on hunting by seal trappers, this species has recovered well. The subspecies in the Galapagos Islands consists of about 30,000 animals. The Japanese subspecies was wiped out.

5. Bottle-nosed dolphin

Population: along the northeast coast of the United States there are about 10,000 to 13,000. In the Gulf of Mexico around 40,000. In the North Pacific, along the Japanese coast there are at least 35,000. In the Mediterranean fewer than 10,000; in the Indian Ocean and along the coast of South Africa, fewer than 1,250. Remainder: unknown. Trend: unknown.

6. Walrus

Population: about 240,000. Trend: unknown.

7. Harp seal

Population: between 2.5 and 4.7 million. Trend: decreasing, but there are still large populations.

8. Sperm whale

Population: about 2 million. Trend: unknown.

9. Hyacinth macaw

Population: 3,000. Trend: decreasing. Deeply threatened by hunting for illegal trade.

10. King penguin

Population: 135,000 to 175,000. Trend: stable.

11. Black rhinoceros

Population: about 2,400. Trend: Following a dramatic 85 percent decline in population (as a result of illegal trade in horns), it is, in

15. Lion
Population: between 30,000 and 100,000. Trend: decreasing, except in protected areas. Especially heavy losses in West Africa.

16. Red deer
Population: about 2 million (worldwide, all subspecies). Trend: increasing.

17. Brown bear
Population (all species) in Europe, North America, and four Asian countries (including Asian Russia): 185,000 to 200,000. Trend: unknown. In Central Europe and North America bears are returning to areas in which they had previously been wiped out.

18. Wolf
Population: between 118,000 and 146,000 (Europe, Asia, and North America combined). Trend: generally decreasing. In China, wolves are hunted as predators. However, in Europe and America they are coming back to areas where they had previously been wiped out.

19. Polar bear
Population: between 22,100 and 27,000. Trend: primarily stable.

20. Giant panda
Population: 1,200. Trend: for over a century, sharply decreasing. Endangered, because the small populations are separated from each other by farms and settlements.

21. Asian elephant
Population: 38,000 to 49,000 (living in the wild). Trend: decreasing (loss of habitat).

general, slightly increasing again. In a few countries there have been further losses, but in South Africa and Namibia, populations are increasing.

12. Chimpanzee
Population: between 105,000 and 200,000. Trend: Some populations are protected and stable, while others are drastically declining (chief cause: hunting for the trade in jungle meat).

13. Gorilla
Population (all subspecies): between 115,000 and 122,000. Trend: same as chimpanzees.

14. African elephant
Population: about 540,000. Trend: After dramatic losses in the 1980s, populations are now nearly stable with a slight tendency toward decline. Significant increases in the countries of southern Africa. A questionnaire sent to large photographic agencies revealed that pictures of elephants are the best-selling animal pictures, followed by pictures of lions.

22. Orangutan
Population: 30,000 to 50,000. Trend: decreasing (as a result of the transformation of tropical forests into plantations).

23. Tiger
Population: 4,600 to 7,200 (all subspecies). Trend: The drastic decline since the 1960s has been halted – at a very low level.

24. Red kangaroo
Population: 9.6 million. Trend: increasing.

25. Koala
Population: between 20,000 and 80,000 (the source for this estimate: Lone Pine Koala Sanctuary). Trend: slightly increasing. Formerly, koalas were hunted for their pelts and nearly exterminated.

Small Animals Shape the Earth

4.5 t 60 km/h

Dung beetles process the dung of mammals. They bury it in the soil and thus reduce the number of flies that use dung as a place to lay their eggs. Many of the 7,000 dung beetle species in the scarab family shape dung into balls. Scarab beetles, which weigh 2 to 5 grams, move balls of dung weighing as much as 244 grams, and attain a speed of up to 20 centimeters per second. This corresponds to a human being rolling a 4.5-ton ball at 60 kilometers per hour.

244 g 20 cm/s

In East Africa, scientists counted 16,000 dung beetles that within two hours carried off 1.5 kilograms of large mammal dung.

Little Animals, Big Effects

Small, inconspicuous animals play the greatest roles in nature. Without tiny krill crabs, enormous whalebone whales would starve. Lions influence the African savannas only marginally, but if all dung beetles were to disappear overnight, the whole ecosystem would suffer severe consequences. While dung beetles work animal droppings into the earth, termites are enriching it with dead plant material. In many areas there are more than 10,000 termites per square meter. Other creatures living in the ground, such as mites and threadworms, are even more numerous. They are all indispensable in shaping the ecosystem and as the foundation of the food chain.

Up to 7.2 million earthworms can live in one hectare of meadowland in Great Britain.

Earthworms are among the most important animals living in the earth. In a year they can carry up to 25 tons of plant material per hectare underground. They transform leaves into humus and aerate the soil. The passageways also serve as tunnels for many other invertebrates. Earthworms are eaten not only by blackbirds and moles, but also, in many areas, by stone martens, badgers, and foxes.

One Square Meter of Meadowland
Number of Animals from Various Groups that Live within a Depth of Thirty Centimeters (selection):

Isopods	200
Ants	400
Spiders	700
Beetles and beetle larvae	900
Diptera larvae	900
Millipedes	1,800
Earthworms	2,000
Oligochaetes	20,000
Springtails	40,000
Mites	120,000
Threadworms (nematodes)	9,000,000

Whalebone whales, as well as many other kinds of fish, seabirds, and seals, feed chiefly on the enormous swarms of krill in the seas. The life development and growth of Antarctic krill is based on a cycle, so the available numbers vary considerably from year to year. In slim krill years, thousands of young penguins and seals die of starvation, while the enormous blue whales leave the Antarctic altogether and seek their food in other areas of the ocean.

The krill population in Antarctica is estimated at up to 400 million tons. Krill is a general term for tiny crabs that live in swarms.

Large Antarctic Species that Feed on Krill
Estimated Consumption in Millions of Tons per Year:

Whales today	33
Whales in 1904	*174*
Penguins	33
Other birds	39
Crab-eating seals	63
Other seals	64
Squid and octopus	100
Total	332

The Ant as a Model of Success

Solenopsis

Orectognathus

Mesostruma

Mayriella

Mystrium

In the Brazilian
rain forest, the dry
weight of all ants
is about four times
greater than that
of all vertebrates
(symbol: the jaguar).

Brazilian
rain forest

The World of Ants
Ants live almost everywhere in
the world. Like human beings,
their evolutionary success is
based on social cooperation.
But whereas all human beings
belong to the biological species
Homo sapiens, ants have diver-
sified into a variety of species.
9,500 species of ants are currently known to scientists (21 of
these are represented here).

90 percent of all
dead animals
(mainly insects)
end up as food in
ants' nests.

Camponotus gigas

10,000,000,000,0

Ants

6,000,0

Human beings

There are about 10,000 trillion
ants, 6 billion human beings,
and 500 ant researchers in
the world.

Ants belonging to the genera
Dinoponera and *Paraponera*
are among the largest. *Camponotus gigas* grows to be
as large as eight centimeters
in length. Ants belonging
to the genera *Solenopsis* and
Leptothorax are particularly
small (c. 0.5 millimeters
in length).

Ants have lived on the earth
for 10 million generations.
Traces of them can be found
going back 100 million years,
into the Cretaceous period.
About 2 million years ago the
development of the human
species began; this corresponds to 100,000 generations.

Strumigenys

Leptothorax

Camponotus

Ectatomma

Polyrhachis

Colobostruma

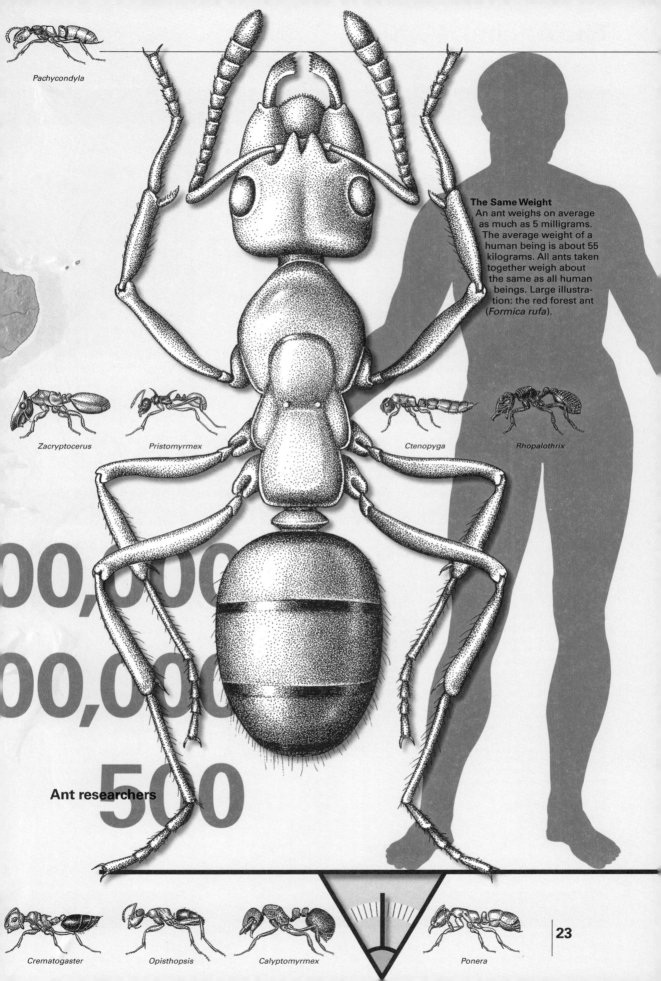

Pachycondyla

The Same Weight
An ant weighs on average as much as 5 milligrams. The average weight of a human being is about 55 kilograms. All ants taken together weigh about the same as all human beings. Large illustration: the red forest ant (*Formica rufa*).

Zacryptocerus

Pristomyrmex

Ctenopyga

Rhopalothrix

00,000

00,000

500

Ant researchers

Crematogaster

Opisthopsis

Calyptomyrmex

Ponera

23

A Changing World

A. *Australopithecus*
B. *Homo habilis*
C. *Homo erectus*
D. *Archaic Homo sapiens*
E. *Neanderthal*
F. *Modern Homo sapiens*

Humans

Tertiary

The Australopithecus learned to walk upright (all chronological indications are approximate)

5 million years   4   3

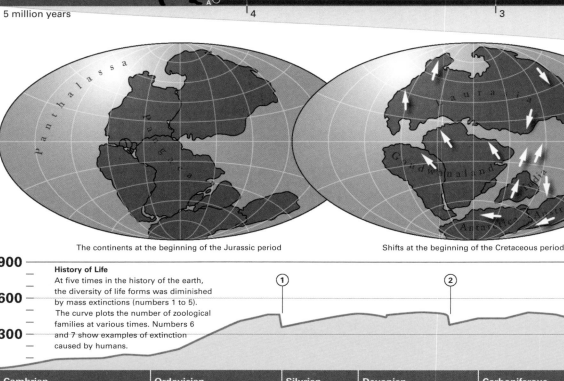

The continents at the beginning of the Jurassic period

Shifts at the beginning of the Cretaceous period

Life

History of Life
At five times in the history of the earth, the diversity of life forms was diminished by mass extinctions (numbers 1 to 5). The curve plots the number of zoological families at various times. Numbers 6 and 7 show examples of extinction caused by humans.

900

600

300

Cambrian	Ordovician	Silurian	Devonian	Carboniferous

Paleozoic
(early age of the earth)

570 million years   500   440   405   350

Nothing Remains as It Is
The number of species that have lived on Earth is estimated at between 100 million and 750 million. In addition to the constant dying out of individual species, there were also repeated climatic and other environmental changes that destroyed thousands of species. Five great mass extinctions led to sharp decreases in the variety of life forms. Afterward, the diversity of life emerged anew. Only 2 to 5 percent of all the life forms that have existed are still extant today. Since humans began hunting animals and altering habitats by the use of fire, they have destroyed species. In the last 450 years alone, humans have wiped out more than 600 known species.

(1) At the end of the Ordovician period, 450 million years ago, 50 percent of all animal families died out, including many trilobites.

(3) 250 million years ago, 50 percent of all animal families died out, including 95 percent of those living in the seas. The Dimetrodon also became extinct in the late Permian period.

In the Devonian period, 30 percent of all animal families died out, including all armor-plated fish. (2)

Earth

Precambrian

4,500 Million Years Ago
Emergence of the earth
Hardening of the earth's crust

3,500
First traces of life
(blue algae)

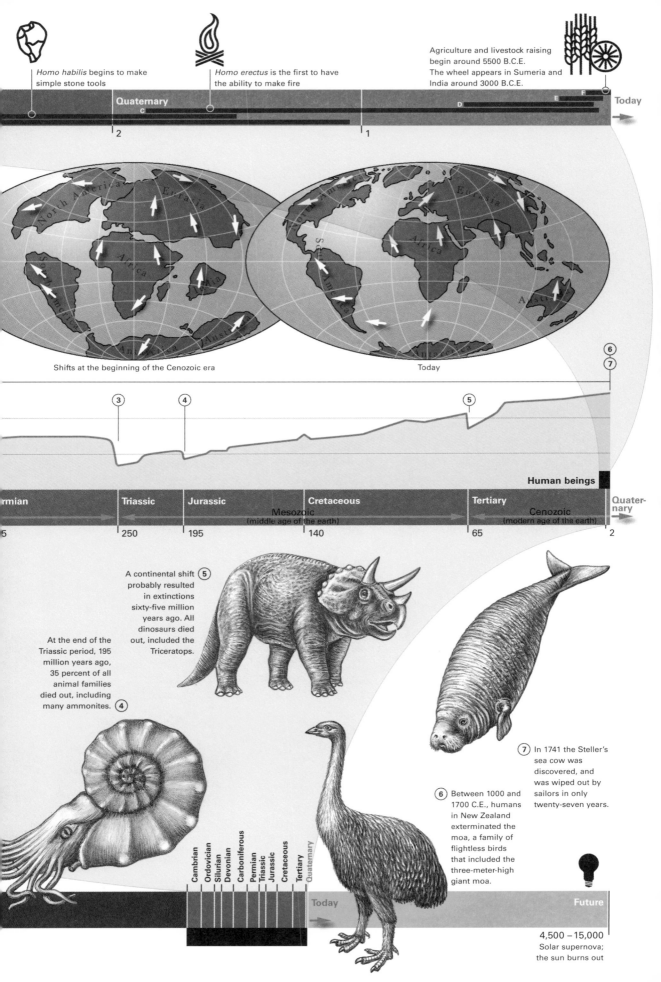

Homo habilis begins to make simple stone tools

Homo erectus is the first to have the ability to make fire

Agriculture and livestock raising begin around 5500 B.C.E. The wheel appears in Sumeria and India around 3000 B.C.E.

Quaternary

Today

Shifts at the beginning of the Cenozoic era

Today

Human beings

Permian | Triassic | Jurassic | Cretaceous | Tertiary | Quaternary

Mesozoic (middle age of the earth)

Cenozoic (modern age of the earth)

250 | 195 | 140 | 65 | 2

A continental shift ⑤ probably resulted in extinctions sixty-five million years ago. All dinosaurs died out, included the Triceratops.

At the end of the Triassic period, 195 million years ago, 35 percent of all animal families died out, including many ammonites. ④

⑦ In 1741 the Steller's sea cow was discovered, and was wiped out by sailors in only twenty-seven years.

⑥ Between 1000 and 1700 C.E., humans in New Zealand exterminated the moa, a family of flightless birds that included the three-meter-high giant moa.

Cambrian
Ordovician
Silurian
Devonian
Carboniferous
Permian
Triassic
Jurassic
Cretaceous
Tertiary
Quaternary

Today

Future

4,500 – 15,000 Solar supernova; the sun burns out

Humans, a Career

Rapid Rise to World Power

Humans and apes parted ways some 5 million years ago. This only seems to be a long time. If a member of each generation were to hold hands with the next, in only 250 kilometers we would be shaking hands with an ape.* Almost 90 percent of a chimpanzee's heritage coincides with that of a modern human being.

*Here we assume that a generation lasts twenty years and that the average arm span is one meter.

Australopithecus

In the course of becoming human, the genus was often threatened with extinction. Here we see *Australopithecus africanus*, one of our various ancestors.

22,000 years ago

500,000 years ago

1,000,000 years ago

35,000 years

60,000 years ago

Highlights of the Process of Becoming Human

The brain of the human genus rapidly grew in size over about 100,000 generations.* Our ancestors lived as hunters and gatherers during 99.6 percent of this time. Agriculture, civilization, and science first emerged in the last 0.4 percent of this time span.

100,000 generations
The length of the development of the brain to its present-day condition

7,500 generations
The length of the development of the ability to use language

2,500 generations ago
The first recording of quantities by using notched sticks

1,000 generations ago
First "worldviews" expressed in cave paintings

250 generations ago
First number symbols used by the Sumerians in Mesopotamia

225 generations ago
Building of Cheops's pyramid

1

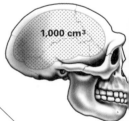

500 cm³   750 cm³   1,000 cm³

Australopithecus africanus

At the beginning of human development 4 million years ago, the "African southern ape," which walked erect, had a brain volume of 500 cubic centimeters.

Homo habilis

About 2 million years ago, the development of the human genus began. *Homo habilis* was able to make the first simple tools.

Homo erectus

About 1.5 million years ago the average brain volume of *Homo erectus* first crossed the one-liter threshold. A completely erect posture increasingly characterized the general appearance of the genus.

106,000,000,000

The total number of all "modern" humans ever born is estimated to be 106 billion. 6 percent of them are now alive.

The Spread of Humans
For four million years, the earliest history of humans took place exclusively in Africa. The Great Trench in East Africa is considered the cradle of humankind. There began the opening up of the world. Ten thousand years ago *Homo sapiens* was already present almost everywhere in the world.

4,000 years ago

14,000 years ago

13,000 years ago

1,600 years ago

years ago

12,000 years ago

1,000 years ago

213 generations ago
First documented population census in China

125 generations ago
Development of the abacus in Greece

40 generations ago
A circle symbolizing "zero" first appears in an Indian manuscript

19 generations ago
First calculating machine invented by Wilhelm Schickard

3 generations ago
Invention of the computer by Konrad Zuse

1 generation ago
Use of microprocessors

Four revolutions in the development of the world's population

1. Local discovery of agriculture 10,000 years ago: increase to 5 million people

2. Worldwide agriculture in 1750: 750 million

3. Massive improvements in health, 1950: longer life expectancy, lower infant mortality: 2.5 billion

4. Change in fertility trends, 1970: trend toward having fewer children. Greater population growth nonetheless, but the speed of the increase in population slows. 3.7 billion

(5.) 1999: 6 billion

1,400 cm³

1,300 cm³

Homo sapiens neanderthaliensis
The Neanderthal was a robust human type specially adapted to cold climate zones. His brain was even larger than that of present-day humans. He died out 30,000 years ago.

Homo sapiens sapiens
With a brain volume of 1300 cubic centimeters, the modern human is currently the last link in the developmental chain.

Humans as Habitat

Leeches live in water and marshes. They make tiny lesions in the skin and drink blood.

The specially adapted mouth of the tick (above, the general appearance of the tick). Its bite can convey life-threatening illnesses.

The bacterium *Escherichia coli* lives in the large intestine of humans. Some varieties can cause severe diarrhea.

Symbiotic and Parasitic Creatures

The human body consists of trillions of cells. However, some of these are not human – they lead their own lives. Some bacteria establish a symbiotic relationship with *Homo sapiens* and help us with digestion and other bodily functions. Others weaken us by causing illnesses in our bodies. Many mites, insects, and worms use human beings as a habitat and source of food, but few of them restrict themselves to humans. All animals and almost all plants have to live with parasites. One-fourth of all animal genera are parasites.

Parasites on Humans
Minimum number of species outside and within the tropics:

■ outside
■ within

	Insects	Mites and Ticks	Threadworms	Tapeworms	Flukes	Single-celled organisms
outside	8	4	12	4	1	13
within	8+	4+	12+	4	6	5

70,000,000,000,

70 trillion bacteria live in a human being's large intestine and aid digestion.

Crab lice live primarily in the pubic hair of human beings. The female lays twenty to thirty eggs which attach themselves to the hair, just as head lice do.

Bedbugs attack human beings while they are sleeping, suck their blood, and then go back to their hiding places.

Itch-mites bore through the horny outer layer of human skin and dig tunnels underneath it. There it feeds on lymphatic fluids and cell contents.

Trichinoids are threadworms that lay their eggs in the small intestine. The larvae nest in the muscle tissue, where they are encapsulated by connective tissue (see illustration). In this state they can survive as long as thirty years.

Head lice live on human head hair, to which they attach their eggs (see illustration above). Lice suck blood and can also convey dangerous diseases, such as murine typhus.

300,000,000
On average, 300 million bacteria live on the skin of an adult human being.

100,000,000
100 million bacteria of 300 species live in the human mouth.

100,000
Only about 100,000 of the bacteria we breathe in every day make their way into the lower respiratory passages.

34

By means of attachment organs on their head ends (here shown greatly magnified), tape worms anchor themselves to the intestinal wall in human beings. They can become several meters long.

Outside the tropics, there are thirty-four kinds of parasites that specialize in humans.

The flea that affects humans is a wingless insect that can make long leaps with its powerful hind legs. Fleas feed on blood. Only the adults live on humans.

Schematic image of the influenza virus, which causes influenza in humans. After each flu epidemic, those who have been infected become immune to a specific subtype of this virus.

Humans as Victims and Prey

Mortal Dangers in Nature
Fauna and flora also have a dark side. Bacteria, viruses, single-celled organisms, worms, and other life forms kill 17,310,000 people per year. Human beings are still far from being able to control nature. Although the battle against smallpox and some other microbes has been won, at the dawn of the twenty-first century many infectious diseases still constitute an everyday danger to human life, especially in developing countries. The threat posed by organisms that cause disease is often underestimated, whereas large predatory animals are a common subject of myths and the media, in complete contrast to the actual menace they represent.

6
humans died in 1998, worldwide, from attacks by sharks. Several million sharks are eaten by human beings every year.

In comparison: death rates not attributable to microbes and other life forms
These figures were calculated at the beginning of the 1990s.

3,000,000
humans die every year from tobacco use.

999,000
humans die every year in traffic accidents.

504,000
humans die every year by drowning.

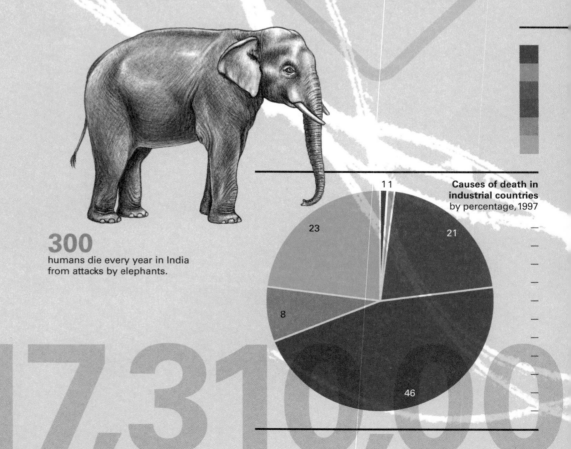

300
humans die every year in India from attacks by elephants.

Causes of death in industrial countries
by percentage, 1997

11

23

21

8

46

SHARKS

SWIMMING

17,310,00

1,500,000

humans die every year, worldwide, from malaria borne
by the Anopheles mosquito
(lowest estimate; highest
estimate, 2,700,000).

265,000

humans die every year
in fires.

56,000

humans die every year from the
effects of alcohol abuse.

The background
shows a salmonella
bacterium, magni-
fied 35,000 times.
This bacterium can
cause life-threaten-
ing diarrhea in
human beings.

- Infections and parasite attacks
- Maternal and infant mortality
- Cancer
- Heart and circulation diseases
- Respiratory disorders
- Other and unknown causes

**The greatest dangers
are invisible. Deaths
from viruses, bacteria,
and other microbes**
worldwide, 1997

Acute inflammation of the lungs	3,745,000
Tuberculosis	2,910,000
Diarrhea	2,455,000
HIV/AIDS	2,300,000
Malaria	1,500,000 – 2,700,000
Measles	960,000
Hepatitis B	605,000
Whooping cough	410,000
Tetanus in infants	275,000
Dengue	140,000

**Causes of death in
developing countries**
by percentage, 1997

- 9
- 5
- 24
- 9
- 10
- 43

2,300,000

humans died in 1997,
worldwide, as a result of
AIDS. The virus originally
came from chimpanzees.
According to one hypothe-
sis, it was transmitted
because hunters were
infected by the blood of
captured apes.

31

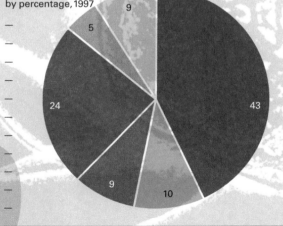

Hot Spots of Biodiversity

950,000

Mexico: Scorpions of the genus *Centurioides* live in the desert.

Mexico
V = 3,045
E = 930
P = 26,071

Colombia
V = 4,764
E = 327
P = 51,220

Ecuador
V = 3,315
E = 312
P = 19,362

Venezuela
V = 3,418
E = 191
P = 21,073

Brazil
V = 6,131
E = 747
P = 56,215

Peru
V = 3,717
E = 340
P = 18,245

Where Nature Is Richest

Biodiversity is not equally distributed. There are hot spots of biodiversity where evolution has produced an especially large number of life-forms. Developing countries in the tropics have the greatest natural treasures. However, people are often the poorest where nature is the richest. Hence, this illustration also shows the gross national product per capita and the population development in these countries. The ten countries with the richest animal and plant life are Mexico, Venezuela, Colombia, Ecuador, Peru, Brazil, Madagascar, Malaysia, Indonesia, and Papua New Guinea. For each country, three indices of biodiversity are given:

V = Total number of vertebrates
E = Endemic vertebrates
P = Total of higher plants

Brazil: Tree frogs of the genus *Dendrobates* secrete a substance that is used by humans to poison arrows.

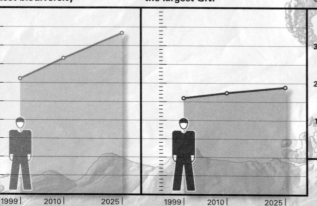

Population growth in the ten countries with the greatest biodiversity

900
800
700
600
500
400
300
200
100

1999   2010   2025

For comparison: in the ten countries with the largest GNP

900
800
700
600
500
400
300
200
100

1999   2010   2025

Per capita GNP in hot-spot and industrial countries
(in U.S. dollars, 1999)

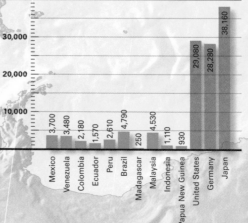

30,000

20,000

10,000

Mexico	3,700
Venezuela	3,480
Colombia	2,180
Ecuador	1,570
Peru	2,610
Brazil	4,790
Madagascar	250
Malaysia	4,530
Indonesia	1,110
Papua New Guinea	930
United States	29,080
Germany	28,280
Japan	38,160

Known species, worldwide
(Selected groups)

Insects 270,000

Plants* 25,000

Fish 25,000

Birds 9,950

Reptiles 7,400

Amphibians 4,950

Mammals 4,630

* Higher plants only

Madagascar: The aye-aye, a primitive primate, can pick larvae out of tree bark with its long middle finger.

Papua New Guinea: The superb bird of paradise lives in mountain forests, and during the mating season it develops a breast shield and head fan made of feathers.

Malaysia
V = 1,865
E = 146
P = 15,500

Indonesia
V = 4,304
E = 1,000
P = 29,375

Papua New Guinea
V = 1,631
E = 323
P = 20,000

Madagascar
V = 1,069
E = 557
P = 9,505

Malaysia

Indonesia

Papua New Guinea

Madagascar

Indonesia and Malaysia: The *Rafflesia arnoldii* produces larger blooms than any other plant. In order to attract insects for pollination, it emits a carrion smell.

Region with greater biodiversity in higher plants

Major production of medicinally significant plants

Ten regions in which endemic bird species are particularly endangered

Twenty-four regions with an especially large number of indigenous plant and animal species

1. Tropical Andes
2. Mediterranean region
3. Madagascar + islands (Indian Ocean)
4. Central American tropical forests
5. Caribbean
6. Upper Burma
7. Atlantic rain forest
8. Philippines
9. Cape floristic province
10. Eastern Himalayas
11. Sunda Islands
12. Cerrado savannas
13. Southwest Australia
14. Polynesia and Micronesia
15. New Caledonia
16. Darien, Chocó, Western Ecuador
17. Western Ghats and Sri Lanka
18. California botanical area
19. Karoo semidesert region
20. New Zealand
21. Central Chile
22. Tropical forests in Guinea, West Africa
23. Wallacea (central Indonesia)
24. Eastern Range, coastal forests

Globalized Nature

Animals as Travelers

Many animal species are extremely mobile. Migratory birds move back and forth between continents. Whales swim from the pole to the equator and back. Some butterflies travel thousands of miles. Even plants do not stay in one place. Carried by the wind or in the plumage of birds, their seeds travel over long distances (see "Traveling Trees," below).

Humans have sometimes unintentionally, and sometimes intentionally, accelerated the spread of plants and animals. They transport all over the world wild game, fish, ornamental birds, or predators that are supposed to destroy pests. Particularly on islands, animals introduced from outside have often destroyed or driven out indigenous species.

Transported by Humans

The animal species represented in silhouette are only a small fraction of those transported all over the world. They all traveled on boats or airplanes. Many of them were intentionally introduced by humans (e.g., starlings in North America), while others escaped from captivity (e.g., the ring-necked parakeet in Germany).

7,300 species living in North America were introduced by humans. 5,000 of these were plant species (or 29 percent of all plant species on the continent), and 2,300 were animal species.

Starling

Wild boar

Muskrat

Ten Important Animal Species and Where They Came From

← from Europe
← from North America
← from Africa
← from Asia
← from Australia

African honeybee

500 [Meters]

hazel

elm

linden

oak

beech

Successful Integration

About 100,000 North American raccoons live in Europe. They are all descended from animals that were introduced in the 1930s. Raccoons have not driven out competitors (martens, badgers, foxes) or exterminated any of the animals on which they prey (young birds, frogs, small mammals).

Traveling Trees

Even without human intervention, nature is not static. Here we see the maximum distance that European tree species can travel in a year by spreading their seed via wind. Birds like the jay can carry plant seeds much farther still.

Ring-necked parakeet

Unsuccessful Integration

Nile perch, which were introduced into Lake Victoria in 1962, destroyed a large proportion of the species unique to the lake. 200 of the original 300 species of fish in the cichlid family have now disappeared or been drastically reduced in number. However, the extent of the ecological disaster in Africa's largest lake is not wholly attributable to the Nile perch; other causes include excessive use of fertilizer on adjoining agricultural lands and a kind of water hyacinth imported from South America that grows over the lake's surface.

1,500

1,000

Rabbit

Red deer

Dromedary

Number of Imported Species in New Zealand
Mammals, birds, plants

39
36
2,000

New Zealand

Brush-tailed possum

Peacock

35

Protected Areas: Four Countries Compared

Increase in Protected Areas Worldwide
Number of wildlife reserves of
more than 10 square kilometers

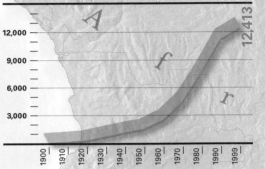

Room for People, Room for Animals
Four countries on four continents:
Two are huge – India and the
United States. Two are middle-
sized – Germany and Tanzania. Two
are densely populated – Germany
and India. Two have large, under-
populated areas – Tanzania and
the United States. Two are rich –
Germany ($28,280 per capita GNP)
and the United States
($29,080 per capita GNP). Two
are rather poor – India ($370 per
capita GNP) and Tanzania
($210 per capita GNP).

In Tanzania, there are, on average,
33 people per square kilometer.

Often, poor countries do more
for the preservation of wildlife
than rich ones do. Tanzania has
put the largest area under protection (the maps show only the
largest protected areas). Even
in densely settled regions, people
and nature can coexist. Take
southern India, for example, in
which there are large jungle
reserves where elephants and
tigers live. Some of the largest
national parks in the United
States are in California.

Large protected areas

Large protected areas
with loosely
defined boundaries

Protected Areas | 28 | 8 | 4.5 | 21 | %

Tanzania: 945,090 km²

* Germany: 356,980 km²

India: 3,287,590 km²

USA: 9,363,520 km²

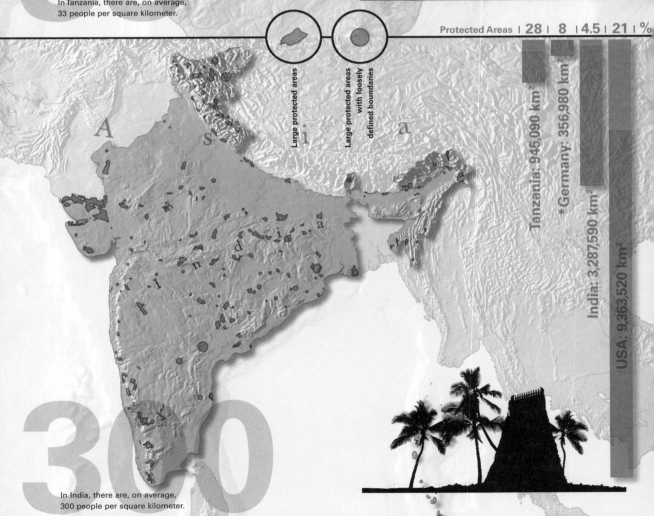

In India, there are, on average,
300 people per square kilometer.

Humanity is becoming increasingly concentrated in cities
City dwellers as a percentage of the total population:
1994 and prognosis for 2025, worldwide

Region	1994	2025
World as a whole	45	61
Africa	34	54
Asia	34	55
Europe	73	83
Latin America	74	85
North America	76	85
Oceania	70	75

* For Germany, more surface area is listed on the UN's list of protected areas. Here, however, only the stronger protected categories were listed, since the others provide a much lower degree of protection. The map shows national parks and biosphere reserves.

30 percent of the earth's surface is land (about 150 million square kilometers). Almost 9 percent is under environmental protection (13 million square kilometers), not counting Antarctica, a world natural park.

10 percent of the earth's surface is covered with permafrost. 40 percent of the ice-free land is used for agriculture, including about one-third as farmland and two-thirds as pasture land.

9   30

10   40

10

%

In the United States, there are, on average, 29 people per square kilometer.

60 percent of humankind is concentrated on 10 percent of the land surface. Settled areas cover 0.36 percent of the continents and islands (500,000 square kilometers).

The protected areas in Alaska are not included on this map.

Survival in Zoos

	Mammals	Birds	Reptiles	Amphibians	Fish
North America	60,000	70,000	25,000	5,000	100,000
Latin America	10,000	25,000	5,000	1,000	25,000
Europe	90,000	130,000	20,000	8,000	180,000
Asia	75,000	100,000	20,000	10,000	50,000
Africa	7,500	15,000	2,500	500	5,000
Australia	7,500	10,000	2,500	500	20,000
Total	250,000	350,000	75,000	25,000	300,000

Estimated World Population of Vertebrates in Zoos: **One Million**

The surface area of all the zoos in the world corresponds to that of one large city. Nonetheless, a great many animals are protected in zoos.

Refuges for Endangered Species
The California condor (large illustration) and many other nearly extinct animal species were saved in zoos. The role of animal parks and botanical gardens fundamentally changed in the second half of the twentieth century. Earlier, they merely exhibited wild animals and exotic plants for money. Today, most zoos and botanical gardens administered by scientists have been transformed into centers for the preservation of species, and in them rare animals and plants are bred. Zoological and botanical gardens can preserve species as long as they are endangered (by poaching, for example) in their natural habitats.

U.S.A.

California

$0.67

Zoos and Visitor Figures

Number of visitors per year, by continent, in millions

Number of zoos per continent

North America 106 / 175

Latin and South America 61 / 125

Europe 300 / 125

Asia 308 / 545

Africa 15 / 25

Australia 6 / 30

Every year, over 600 million people visit zoos – that is, one out of every ten people, worldwide.

Six Leading Botanical Gardens and the Number of Plant Species They Contain
1. Royal Botanic Gardens, London (Kew), 34,000 species
2. Berlin Botanical Garden, 22,000 species
3. Royal Botanic Garden, Edinburgh, 17,000 species
4. Botanical Garden, New York, 15,000 species
5. Botanical Garden, Munich, 14,000 species
6. Palm Garden, Frankfurt, 13,000 species

619,000,000
people per year visit zoos.

1,000,000
vertebrates are housed in zoos throughout the world.

80,000
plant species live in botanical gardens worldwide.

7,000
animal species are housed and bred in the world's zoos.

2,000
species of land vertebrates, maximum, can be saved in zoos.

1,800
botanical gardens exist worldwide.

1,200
zoos exist worldwide and are organized into associations.

30
species of animals that have become extinct in the wild now exist only in zoos.

The California Condor
If we divide the amount that is spent on saving this bird by the number of people living in California, the result is sixty-seven cents. It was once shot and poisoned as an alleged livestock predator. Conservationists captured the last remaining birds, bred them in zoos, and reintroduced them to their natural habitat.

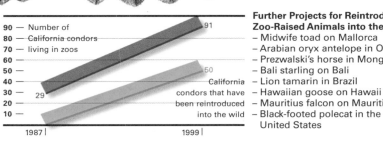

90 — Number of
80 — California condors
70 — living in zoos
60 —
50 —
40 —
30 — 29
20 —
10 —

91

150

California condors that have been reintroduced into the wild

1987 | 1999 |

Further Projects for Reintroducing Zoo-Raised Animals into the Wild
- Midwife toad on Mallorca
- Arabian oryx antelope in Oman
- Prezwalski's horse in Mongolia
- Bali starling on Bali
- Lion tamarin in Brazil
- Hawaiian goose on Hawaii
- Mauritius falcon on Mauritius
- Black-footed polecat in the United States

Animal Censuses in the Serengeti

A Worldwide Network
Approximately 200,000 people count animals worldwide: zoologists and conservationists, gamekeepers and hunters. This global monitoring of animal populations provides information on the state of nature and gradual ecological changes. At the WCMC (World Conservation Monitoring Centre), which produced most of the information for *Life Counts,* the results were collected and evaluated.

The Serengeti
Since 1957, when Bernhard and Michael Grzimek began to count animals from the air, the steppe animals in the Serengeti plain in northern Tanzania have been regularly recorded. No other animal population has been counted for a longer time. Year after year, the herds follow the rain clouds in order to find fresh green grass. In 1998, the approximately 920,000 white-bearded gnus, 320,000 Thomson's gazelles, and 150,000 Grant or plains zebras traveled 2,200 kilometers.

In 1986 the Serengeti experienced a plague of caterpillars (see text at left below). End to end, these "army worms" would have reached around the equator 222 times.

222

With a population of about 920,000 animals, the white-bearded gnu is one of the most common cloven-hoofed animals in the Serengeti.

Who Is Heaviest?
Taken together, all large animals in the Serengeti weigh less than the region's population of one species of owlet moth caterpillar (*Spodoptera sp.*), which in many years appears in enormous numbers. During the caterpillar plague of 1986, the creeping horde weighed, according to the highest estimates, over 320,000 tons. However, the sum weight of the Serengeti's grass is still the heaviest: Its total biomass amounts to 5.6 million tons.

323,000

Millions of tons

6
5
4
3
2
1

Grass biomass	5,600,000
Caterpillar biomass 1986	392,000
Median biomass of all large animals	313,000
Hoofed animal dung, per year	455,000

Changes in the Population of Gnus, Thomson's Gazelles, and Zebras
In millions, 1957 to 1998

- Gnu
- Thomson's gazelle
- Zebra

1.5

1.0

0.5

1955   1960   1965   1970   1975   1980   1985   1990   1995   2000

T a n z a n i a

41

Statisticians on Safari: Our Methods for Counting the World

Every year, 10,000 biologists set out to determine the animal population worldwide. The statistics they assemble serve research and conservation. However, even the largest animals, such as elephants, are hard to count.

Four hundred and ninety-six gnus – click. Thirty gnus – click. One hundred and thirty-three gnus – click. In the belly of the airplane, Sarah takes pictures at regular intervals. Down below, on the southern plains of the Serengeti, 10,000 white-bearded gnus are moving and grazing in an apparently random way. Above, at an altitude of 150 meters, Sebastian, the pilot, is flying the black-and-white striped Cessna over the savanna, following a grid of invisible but very precisely laid-out lines. His systematic crisscrossing ensures that it will be possible to count the gnus in the photos in a statistically correct manner. Three hundred and eighty-six gnus – click.

Below, on a grassy knoll fifty meters high, the wilderness tries to keep itself from being turned into statistics. Now the gnus, which look like a cross between a buffalo and a goat, seem to be preparing to take a leisurely break. Suddenly, individual animals rush ahead, larger groups start to move; then the whole herd takes off. Why this sudden departure? Bleating, grunting, and growling, the herd moves through the greenish-yellow shimmering sea of grass, with bobbing heads and swinging tails, slowing to a trot, then galloping again. When are they moving slowly, when quickly? Dusty, branching trails that diverge, come back together, and then drift apart again serve as the path to rain – yet then one group of bulls suddenly turns west. Why here, and for how long? The two people in the plane can observe individual changes of direction, but not explain them. Effects without causes. Chaos instead of coordination. Gnus know nothing about geometry; turbulence and spontaneity determine their course. On these plains, a single hoofbeat can provoke a flurry of movement.

But the count must be made. Without exact data indicating whether the herds of large grass-eaters are growing or shrinking, humans cannot make efforts to protect them. Therefore we must try to translate the welter of movements into clear tables. Sarah and Sebastian work for the Frankfurt Zoological Society (FZS), a private conservation organization that regularly surveys the wild animal population in Tanzania. That is why the plane with the insignia "5H-Zoo" is flying over the plains; that is why Sarah is clicking the shutter of her camera every twenty seconds; and that is why Sebastian is flying straight along the grid lines. He is following the coordinates sent by the satellite-based Global Positioning System to a small receiver that records them on paper. He follows a straight course from north to south, makes a tight turn to the east, then goes back. Scientists call this grid of precisely defined flight lines a transect, along which the animals are counted. Each transect is a kind of random sample of the whole area, which is enormous. In the language of the Masai, "Serengeti" means "endless expanse."

In the cockpit, there is nothing of the romantic conservationist in the "Out of Africa" style. Instead, sober, clockwork precision is practiced. Hour after hour, the census flight goes on. The motor drones, drowning out all other noises. After three quarters of an hour, Sarah puts in fresh film. After four hours, Sebastian kills a tsetse fly; this will not appear in the subsequent record. Every two or three years, specially trained scientists and rangers of the Tanzanian National Park make an aerial survey of an area of 250,000 square kilometers – the size of Great Britain. During the major census, everything in the animal realm that can be clearly seen from the air will be counted: buffalo and elephants, lions and zebras, giraffes, gnus, and gazelles. A laborious undertaking. And an expensive one. But Dr. Markus Borner, the FZS's representative for East Africa, is convinced that this is a useful activity. "The animal population and its development are the basis of all our planning – in the battle against poaching, for example, one of our biggest problems. The aerial censuses tell us where the losses are especially high." Then ranger posts are set up or reinforced at these places. Or the park administration and the FZS set up buffer zones along the boundaries of the preserve, so that the migrations of the white-bearded gnus – a key factor in the economy of the Serengeti – can also move along secure paths through the national park.

"We need statistics," Dr. Borner says, "in order to gauge the success or failure of our efforts to protect the animals." One failure is reflected in the fact that the remaining population of rhinoceroses in the Serengeti can be counted on the fingers of one hand. One success is the increasing number of elephants in the large Selous preserve in Tanzania, which shows that the ecosystem is in good condition there. "Poachers had reduced the population of elephants from an original 110,000 to 20,000. But since the early 1990s the herds have recovered." Both the prohibition on trading in ivory and the practice of sharing hunting proceeds with surrounding villages have contributed to this success.

But can the enormous herds of the Serengeti really be counted? Did exactly 923,460 white-bearded gnus graze on the endless plains in 1998, as reports claim? Borner admits that this figure cannot be precise but adds that it does not need to be: "We are interested less in absolute numbers than in large-scale trends." The Systematic Reconnaissance Flights (SRF) were developed in order to produce informative statistical curves. These offer a standardized method for providing scientifically tenable data.

Handheld cameras and video cameras have been used in SRFs only for a short time. In the classic method, the animals are counted by human observers. The observer sits next to the pilot and notes down all the landmarks visible on the ground: bodies of water, fire sources, hills, vegetation, possibly also the marks of human habitation such as houses and fields. These observations later play a role in evaluating the data. Sitting behind in the plane (most are small, single-engine planes) are two people who do the actual counting. The field of vision within which they record the animals is marked out by two horizontal rods attached to the plane's wings, in such a way that the counter (assuming the altitude is correct) views a strip of ground of precisely defined breadth, usually 150 or 200 meters. Only animals within these limits are counted. So that the observers can keep their eyes constantly on the strip to be counted, they do not write down the numbers but record them on tape: "Zebra, zebra 35." So far, so good.

But a gremlin is always onboard. An imprecise distance from the ground can throw off the results, and so can a counter's unacknowledged shortsightedness – not to mention the possibility of air sickness. Scientists compensate for the inadequacies of human beings – who, after all, were not born to be computers – by using an arsenal of mathematical formulas. Not every counter is equally capable of counting thirty-five zebras in a short time or of recognizing them from a considerable height. However, using statistical data drawn from previous experience, scientists decided that the figures given by the observers seated on the right and left should not deviate too much from a certain average value. If they do, then it is suspected that, for example, one of the two observers is regularly counting gnus as zebras, and additional hours of in-flight training are prescribed. Another procedure ensures that the views from the right and left sides of the plane always cover the same defined breadth. To this end, at the beginning of a census flight the pilot repeatedly crosses a row of marked stones. Only when both observers count the same correct number of stones is the field of vision considered to be "standardized."

When the plane is back on solid ground, a computer is used to perform a series of complicated calculations on the manually or photographically collected data. Since no one is able (or willing to spend the necessary money) to completely survey an area as large as the Serengeti, transects are counted and then extrapolated to the whole area. "In doing so, we take the standard statistical margin of error into account and thus can guarantee that they are ninety-five percent accurate," Dr. Borner says. He suggests 923,460 as a fairly accurate estimate of the total number of white-bearded gnus in the Serengeti – though this number could be off in either direction by as much as 198,959. In any case, it is relatively large, and certainly at least five times as great as it was at the time that Bernhard Grzimek was rightly concerned about the Serengeti's ability to survive. Grzimek, who was then director of the Frankfurt zoo and a well-known filmmaker, visited the East African savanna toward the end of the 1950s to observe the movements of the great herds. In the Serengeti, he discovered two things: first, gaping holes in our knowledge of one of the most fascinating animal paradises in the world, and second, plans to drastically reduce the size of the preserve, despite the fact that it was so little known. Grzimek believed he could use hard facts to put gentle pressure on the politicians responsible. So he decided to count up "this swarming anthill called the Serengeti." But how? His son Michael answered the question: "We just have to learn to fly."

Father and son had hardly received their pilot's licenses before they developed their method of reconnaissance flights using transects. This method was later refined by other scientists at the Serengeti Research Institute, and today it is used throughout the world. From the first, it was very difficult to fly, observe, and count. "We started mumbling numbers in our sleep." After the Grzimeks had flown up and down the Serengeti, they had counted a total of exactly 366,980 animals – chiefly the highly visible grass eaters. Hardly anyone in the world had ever counted that large a collection of animals. (Later investigations with improved instruments showed that the herds were in fact considerably larger.) Although the migrations of the herds are unpredictable in detail, they are nonetheless regular enough when taken as a whole. Each year hundreds of thousands of white-bearded gnus in the southern Serengeti start moving north, following the threatening rain clouds as a dark promise. Rain means grass, and grass means life. The Grzimeks discovered that the traditional migration routes led far beyond the existing park boundaries, but outside the protected area, gnus and zebras ran straight into the traps set for them by poachers. Both Grzimeks were not only good counters but also good dramatists, and they documented their discoveries in a film called *The Serengeti Must Not Die* (*Serengeti darf nicht sterben*). The film weaves science and adventure together in a fascinating way. It enjoyed worldwide success and won an Oscar, and the Serengeti reservation was not reduced but rather was significantly extended.

The East African savanna is a stroke of luck for science and conservation. Because the Grzimeks began to count forty years ago, we have long-term data. The usual case is quite different. In most regions of the earth there are more gaps than knowledge. Corresponding to the unequal distribution of research communities around the world, most of the available figures concern the temperate latitudes, whereas the tropics are uncharted. No one knows exactly how many elephants live in Africa, how many dolphins swim in the ocean, how many jaguars roam the South American forests, or how many pythons slither through the Indian jungles. But such figures are indispensable for providing a foundation for the protection of animal populations.

The most important reason for the lack of illuminating or even vague figures for many areas is, as so often happens, money: Somebody has to pay for the counting. Environmental monitoring is expensive. Even if the means are available, the path to good data is full of obstacles. Censuses are particularly difficult to obtain:
– access to the area may be difficult (as in the tropical rain forests);
– the animals in question may be nocturnal (such as lemurs or owls);
– the animals in question may be well camouflaged (as are snow leopards);
– the area may be under water (as in the case of fish on reefs) ;
– the species may, under natural conditions, have only a few individuals spread over a large area (such as tigers);
– the species may travel long distances (like sea turtles). For such populations, there are either only questionable data (with a tendency toward undercounting) or only rough estimates. Or, as in most cases, there are no data at all: The animals boycott the charts.

We know most about the species with which we have a gut-level relationship – that is, the ones we eat. So the populations of species that are hunted, such as the European roe deer, have been well known for centuries (even though they have been mainly underestimated). The same can be said about the development of the North American bison since 1700 or the number of fur seals along the Namibian coast. Research has developed its own formulas for determining the general population on the basis of the kill made by hunters. In the case of the roe deer, for instance, one multiplies the kill by three; in that of chamois, by five. The differing multipliers reflect the fact that the size of hunting reserves, birthrates, mortality, etc., may vary.

With fish, too, we tend to research what we like to eat. Developments relating to food fish are regularly analyzed. Here, too, the size of the catch is extrapolated to the total population. The International Council for Sea Research (ICES) provides annual statistics on one hundred species fished in the North Atlantic, and has done so for a long time so that future trends can be relatively easy to predict on the basis of the size of the catch, mortality, birthrates, and biomass. However, this assumes that the figures that ship captains write in the logbooks regarding the size of their catches are accurate – and this cannot always be assumed. Statisticians have methods for correcting such imprecisions, but these methods sometimes produce their own errors.

In rare cases fish are even counted individually. In coral reefs, for example, researchers are interested in biodiversity or biological productivity: How many fish does a reef produce; how many can it feed? Underwater, methods are used that have already proven their worth in the Serengeti. Before diving, the researchers establish a transect as the area for study. This line is marked by a rope stretched across the sea floor. Two divers swim along the rope – one on the right, one on the left – and count the fish they see. To be sure, as British marine biologist Mark Spalding says, "One can concentrate on only one or two dozen species at a time. You can't do more than that; otherwise you'd go crazy in the midst of the colorful teeming throng." In addition, divers have to cover the transect in the shortest possible time; otherwise they run the risk of counting the same school of fish twice if in the interim it moves farther on in the same direction. (This, by the way, is a trick well known to makers of Westerns: They have the Indians cross the camera several times, giving the spectator the impression that there are thousands of them.)

Solely in the interest of offering a complete account, Spalding mentions the one truly exact method of counting fish: a couple of sticks of dynamite. This method is seldom practiced, but is dead accurate in two respects. Just the right amount of explosive is used, so that when it is ignited in the water, it causes the swim bladders of all the fish in a circumscribed area to burst. The fish then float to the surface and can easily be counted. The problem, of course, is that conservationists refuse to count a population to death, and there is a methodological disadvantage: The census is no longer repeatable.

For each group of animals, biologists have developed a special technique ranging from life counts to death counts:
– Insects, especially moths, can be collected at night using light traps.
– To reach the beetles in a canopy of the rain forest, trees are sprayed with an insecticide. Shortly afterward, the insects fall into nets spread underneath the trees.
– In open country, predators such as lions and leopards are counted from all-terrain vehicles, following a set of parallel lines.
– Birds are counted by having observers move along transects and note down all sightings.

The counters themselves are counted, too. It is estimated that, worldwide, more than 200,000 people take part in such censuses every year. Among these are the more than 3,000 scientists and conservationists of the Declining Amphibians Population Task Force, who document amphibian populations in ninety countries. And the members of the International Waterbird Census. And the more than 100,000 enthusiasts who participate in the NTT World Bird Census. Their annual activities combine ornithological sensitivity with athletic competition. Their goal is to observe as many bird species and individuals as possible. Each reported species earns money for conservation: The Japanese telecommunications company, NTT, spends about forty U.S. dollars per species on an international bird preserve.

Perhaps the oddest kind of wildlife census is found in the British Isles. In a country where people who otherwise seem to be in full possession of their faculties indulge in hobbies such as "trainspotting" – hunting for serial numbers on locomotives passing by – it may not be surprising that we also find "twitchers." Why they are called "twitchers" remains obscure, but we do know that the term refers to members a secret society of bird-watchers and counters who call a special telephone switchboard every time they observe a rare species of bird. The rarer the species, the more excited the group becomes. In a single hour on a single highway in Scotland, the police once stopped ninety drivers for speeding: They were all hurrying from various parts of an island to where a gray-tailed water runner had been spotted. Twitcher Lee Evans was behaving typically when he left his own wedding because an Isabel shrike had been sighted somewhere nearby. The marriage was later completed, but did not survive the first bird-watching season.

Somewhat more methodically, but with no less commitment, scientists all over the world set out every year to inventory animal populations. They collect frogs in Australia, track caribou along the Yukon River in Canada, and record monarch butterflies in North America. They bring science to bear on the wilderness by sheer force of will. On one hand, theirs is the intellectual world of the researcher, shaped by models and formulas, fired by the desire for order, mindful of discipline; on the other, it is the factual (animal) world, full of chance and chaos, tending toward unpredictability. Between them are numbers, which we have employed as neutral mediators in order to explain nature; to conceive the world, at least in numerical form. They represent contexts that are in reality far more complex than any series of numbers can express. Numbers are willing simplifiers. That is why we are so eager to count things.

One might think that counting elephants would be particularly uncomplicated. After all, they are seven meters long, weigh several tons, and move at a rather leisurely pace. Yet they show how errors in counting can multiply – and ultimately even make political careers. In 1979, the World Conservation Union (IUCN) organized a continent-wide elephant census in Africa. It sent questionnaires mainly to researchers, national park rangers, and various animal experts. Altogether, 1.3 million specimens of *Loxodonta africana* were reported. When ten years later another group of experts once again set out to count all the elephants in Africa, the result was shocking: Only 625,000 animals were left. Half of the African elephants, it seemed, had disappeared within a single decade, hunted down for their ivory, slaughtered by unscrupulous poachers. The discovery provoked an international outcry of dismay. At the next international conference on trade in endangered species, the sale of ivory was prohibited throughout the world. Photojournalists reported on bloody slaughters; a deeply moved public spent millions; and private conservation groups and government donors sent large sums of money to Africa. Jonathan Adams, who was then working for the World Wide Fund for Nature (WWF), recalled that "a regular elephant-mania erupted."

At the same time, doubts arose. Was the census made ten years earlier conducted in a rigorous manner? In their book, *The Myth of Wild Africa*, Adams and his colleague Thomas McShane described how scientists reexamined the questionnaires and found inconsistencies. At the time of the census, Niger was suffering from a terrible drought, and rangers had other things to do besides count elephants. In Malawi's Kasungu National Park, one observer constantly slipped up during survey flights and counted all the animals outside the marked field. In nineteen of the thirty-five countries involved, the results were based on a single source of information, without any way of checking it. One counter was even responsible for all the elephants in five countries. The examiners concluded that the 1979 report probably listed more animals than actually existed. The world mourned the death of hundreds of thousands of elephants that had never lived. Corrections of the census were published, but only in scientific journals.

And this is why the general public still tends to think of African elephants as being endangered. In reality, recent surveys show that the development of the elephant population varies with the region: Whereas in parts of East Africa they continue to be threatened by poachers, in southern Africa they have so splendidly increased that even conservationists are recommending limited hunting with profits from the sale of the ivory going toward conservation.

The conflict shows how important regular censuses of life-forms are. This is particularly true when a species of animal is managed. Only repeated surveys can show whether a population is being used in a sustainable way, or whether, instead, more are being taken than are being reproduced. The dynamics of populations have to be documented over many years. Statistics are like photos: One cannot see changes in a snapshot, but only in a long series of photos taken over time.

Species of animals that are not used commercially must also be observed. Monitoring serves as an early warning system for protecting species: Where are the populations decreasing? What is threatening them? We know how the last surviving populations of Siberian tigers were faring at the end of the 1990s only because 675 counters working for the WWF spent a whole winter fighting their way through the Russian wilderness in order to survey several thousand tiger trails. Big cats require enormous reserves, and for this reason they seldom live under natural conditions. The conclusion drawn by the observers after completing their frosty labors was that although poachers were still going after them, their population seemed to have stabilized at a low level of between 415 and 475 animals.

The "leg-counters" among biologists are almost inevitably overshadowed by those who are involved in more spectacular conservation activities. However, the environment needs not only committed Rainbow Warriors but also committed bookkeepers. In 1998, this insight led the WWF, the IUCN, and the United Nations Environmental Programme to cooperate in the establishment of an international center for conservation censuses. Since then, streams of data have flowed into the WCMC as through into a global brain: The organization uses the numerous counters in the field as its eyes and ears, and the Internet as its neural conduits, while analysts using computers serve as its central nervous system.

Disappointment awaits anyone looking among the rape fields and idyllic lanes on the outskirts of Cambridge for the kind of building that multinational organizations like to construct for themselves. The WCMC's modest two-story headquarters, which is built not of marble but of brick, is hidden in the parklike English countryside. All over it – in work spaces, in halls, next to the coffeemakers – there are stacks of professional journals, large-format volumes on plant identification, note papers, unrolled maps, and letters bearing stamps from all over the world. Many of the fifty or so members of the center's staff – arctic experts, freshwater researchers, tropical botanists, marine biologists – prefer sturdy shoes and khakis, as though they are preparing for an expedition into the jungle. In reality, the wilds lie far away. "Unfortunately, we seldom get to do fieldwork. Most of the time, we evaluate information provided by others," says Dr. Brian Groombridge, the coordinator of the Biodiversity Section (and director of scientific research for this book). The WCMC bases its analyses on the work of a worldwide group of scientists who deliver firsthand data to Cambridge. On this basis, voluminous works such as *The Red List of Endangered Plants*, the definitive *Global Biodiversity*, and a world atlas of mangrove forests endangered by shrimp farming are produced. Earlier, we could only guess at the extent of the damage like that done to the mangroves. "We ourselves are often astonished," Groombridge says, "to see how little nature in a region is known before we begin our research there." He and his colleagues are producing new knowledge by collecting scattered information, analyzing it, and then making it available in the form of data banks and maps to both experts and lay readers. This unique combination of meticulous scientific work and popular presentation made the WCMC an ideal partner for the Life Counts Project.

The motto of the Cambridge master counters is: "There is only one living earth, and so we count it." Beyond this kind of emotional appeal, the work routine in the WCMC is like patiently brooding over a 50,000-piece jigsaw puzzle in the hope that someone will ultimately be interested in the complete panorama. And none of these panoramas lasts for long. "Every new expedition can completely change our picture of a biotope," Groombridge says. Nature and human beings are always mixing up the pieces of the puzzle; then they have to be laid out anew.

Statistical surveys of plant life are especially expensive. In contrast to animals – where the problem is to get close enough to count them – in the case of plants the researcher has to take a step back in order to see them better; vegetation is best viewed on a wide scale from satellites. Long-distance reconnaissance is the generic term for this program of mapping from space everything covering the earth: moors and savannas, rain forests and timber plantations, deserts and pastures. Satellites make use of the fact that every growing thing emits a specific type of electromagnetic radiation, consisting of reflections from the sun or from its own heat. Sensors on satellites pick up these impulses. They use the spectrum between ultraviolet and infrared to tell us, with an accuracy down to five meters, which plants are growing 900 kilometers below them.

For the rain forest, surveys from space are a real boon. Long-distance reconnaissance is always called for when studying large biotopes to which access is difficult. The tropics are not only a hot spot of biodiversity, but also a place where it is particularly endangered. Here, every acre counts. A satellite map shows jungle all around the Brazilian city of Manaus. A computer translates each type of electromagnetic radiation into a different color: The untouched areas, the so-called primeval forest, are dark green; the Manaus metropolitan area, whose suburbs are constantly nibbling away the jungle, is pink; the Negro river, which at this point joins the Solimões river to form the Amazon proper, is black. A superhighway is also visible, leading north, from which other, smaller roads branch off, all of them surrounded by innocent-looking light green. The latter, however, is the color that indicates areas that have been cleared of vegetation. The computer image tells a whole story. It begins with the loggers who use bulldozers to cut corridors into the rain forest, and ends with settlers seeking gold, rubber, and their fortunes in the areas thus opened up. In the tropics the story varies little. The annual loss of rain forest in the Amazon alone amounts to 25,000 to 30,000 square kilometers, an area the size of Belgium. But there is no figure that is unchallenged by someone. Some scientists estimate that the loss of rain forest is much greater than the satellite surveys indicate. On logged-off areas, they argue, plants grow back which no sensor can distinguish from untouched forest. The debate shows the importance of parallel monitoring on the ground in order to confirm or disconfirm the results of long-distance reconnaissance.

Around two dozen satellites circle the earth in order to measure surface-level temperatures, ocean currents, and the growth and shrinking of glaciers, and to locate storm fronts and cyclones. The United States, Europe, and Japan are not the only participants in these programs; poorer countries such as Russia, Brazil, and India have also launched their own satellites into orbit – even though they cost more than $50 billion apiece (for rockets, rocket transport, launching pads, and service over ten years).

But this kind of ecological early warning system can save the harvests in whole regions. For instance, one of African farmers' dangerous enemies is *Locusta migratoria*. Migrating locusts descend in swarms that may include between several million and several billion insects. Depending on which way the wind is blowing, they can fly as far as 2,000 kilometers. Where will they land next? Environmental monitoring can help answer this anxious question. First, insect experts observe the locusts' breeding grounds and migration routes. This information is combined with satellite data on temperatures, rainfall, and vegetation growth. Then farmers are advised to harvest their crops before the swarms of locusts come and, in next to no time, eat up the whole countryside.

The observation of locust migrations is part of a program the United Nations Food and Agriculture Organization (FAO) has set up for arid regions such as the Sahel. With regard to failed harvests and locust attacks, environmental monitoring is no longer something for conservationists and fact-loving researchers alone. Existential motives move into the foreground: Count lives in order to save lives.

See also plates:

Human Favorites
Page 18

Animal Censuses in the Serengeti
Page 40

45-million-year-old fossil of a turtle from the Eocene, North America

Fossils: Nature's Methods for Counting the World

For by far the longest period in the history of the world, nature was its own chronicler. Like the hard disk of a computer, the fossilized traces of earlier life contain enormous amounts of information that we can call up today.

Stegosaurus

Two powerful legs slowly plod through the silt. The heavy animal, which weighs several tons, carefully moves forward, raising and lowering its head like a hen. Three toes imprint themselves on the soft ground; they are half a meter long and have sharp claws in front. The length of its stride is about two meters. The enormous reptile's tail swings slowly and threateningly back and forth. Its mouth is slightly open, and its breath stinks. The animal is the spitting image of a tyrannosaurus. It suspiciously surveys the area and leaves behind deep footprints.

This scene took place 135 million years ago, on the edge of what is now Australia's Kimberley Region. There, on Gantheaume Point, near the small coastal village of Broome, stands a lighthouse. At the foot, an amateur paleontologist came across fossils in the red cliffs. On that day, the ebb tide had withdrawn to an unusual distance and revealed to view depressions that turned out to be the footprints of *Megalosaurus broomensis*. Farther down, where the whitecaps were breaking on the azure blue Sea of Timor, a river valley once meandered. After the last ice age it was submerged by the rising ocean.

The earth has circled the sun 135 million times since the prehistoric animal's footprints were made, and the moon has made twelve times that number of rotations around the earth, but we have only recently gotten on the merry-go-round. The life span of our species is too short to be able really to grasp the huge scale of geologic time. Dinosaurs ruled the planet more than 1,400 times as long as humans have. For the Kimberley Region, 135 million years is not a particularly long time. Stone fragments from the high plateau in the northwest corner of Australia have been dated at 1.9 billion years old. In other places in Australia, stones up to 3.5 billion years old have been found that contain more than eleven different kinds of identifiable fossil bacteria.

For over 99.99 percent of the earth's history, nature inventoried its creations by itself – from the smallest living creature to the most enormous reptile – before humans made their first drawings on cave walls. Like a computer's hard disks, fossil remains of earlier life contain enormous amounts of information that we can call up today. Here, a footprint or a mussel on the stone, there an insect caught in amber or a mammoth preserved in ice.

Rhamphorhynchus

Sometimes the planet has carefully laid down the fossils of manifold life in clearly distinct layers, but sometimes these deposits are all mixed up with each other or preserved in a slipshod way. However, most of them have been capriciously destroyed and subsequently recycled. It is not for nothing that the coal we burn in our furnaces and the gasoline we put in our cars are called fossil fuels. They were produced by septillions of plants and animals that once lived on the planet. The figure septillion is chosen here only because we can still somehow describe it: It is a one followed by twenty-four zeros. Thus we find ourselves back in eternity: 157 quadrillion seconds ago – that is, 5 billion years ago – the earth did not yet exist. And to get from 157 quadrillion to a septillion, we still have to add seven zeros.

When we try to make a quantitative balance sheet of past life, time and numbers both approach infinity. How could it be otherwise, if we consider that a single bee population consists of 60,000 to 70,000 individu-

In 100 million years, what fossils of present-day human beings and their culture will remain? If the level of the oceans rises, the preservation of such remains will become more probable. Tectonically sinking cities such as Amsterdam, New Orleans, Cairo, and Venice are more likely to leave traces behind than are rising cities such as Los Angeles. Bricks, soft drink bottles, and objects made of plastic could be fossilized. [1]

als, and a large population of forest ants has about a million members? The largest documented swarm of locusts struck Nebraska in 1874 and was estimated to include several trillion individuals, which won it an entry in the *Guinness Book of World Records*. The number of individuals can be digested at most in small amounts, and even the fossil chunks are sometimes pretty big: The American fossil hunter John Horner discovered in Montana traces of a reptile herd consisting of some 10,000 animals.

As attractive as the idea of a numerical census of individuals of past species might be, it far exceeds the capabilities of science, even in the area of methodology alone: Which species do we want to count? Ultimately, we have to conceive of a species as a changing envelope, like variations on a theme. But even if the researcher proceeds more modestly and seeks to determine how many different species have lived on Earth, he or she will quickly realize that the task would take an eternity to complete.

Coelophysis

Nature has preserved a record that has some gaps but nonetheless remains enormous; in practically every corner of the planet, "exhibit cases" underground hold fossils awaiting discovery and classification. Gantheaume Point in the Kimberley Region is one of the magical places where humans can get a glimpse of what really counts. The visitor ends up feeling humble when confronted by nature operating on such a gigantic scale. Towering cliffs rise out of the sea, terrifying crevices disappear into dark depths, and former coral reefs poke through the red sand like huge mountain tops. And we must not forget the weather, with its biblical extremes, ranging from deluges of rain to simmering heat. The largely unexplored landscape of the Kimberley high plains basically looks like one huge fossil with primordial thunder and lightning. This is "Jurassic Park" live.

Nonetheless, in this archaic landscape we find some living fossils as well. A fisherman who has seen the serrated tail of a saltwater crocodile (*C. porosus*) pass by his rowboat could think he has suddenly slipped into the Mesozoic era. At up to seven meters long, this reptile of the Kimberley was part of the dinosaur family in the Mesozoic and has hardly changed since. Crocodiles were put under protection in Australia and have survived astonishingly well. By 1945, they had almost been wiped out, but today the population of these living fossils is doing quite well: Tens of thousands of "salties" lurk in the salt water along the coast and in the mouths of tropical rivers, looking for prey. Their population is now more or less what it was before they began being exterminated. Only individual crocodiles with bad habits, such as eating careless tourists for lunch, are hunted down or resettled.

On the stage of life many actors appear only briefly, while others remain for an amazingly long time. The life expectancy of a genus of mussels averages about 80 million years, while that of a genus of brachiopods about 20 million years, and that of a genus of ammonites about 8 million years. A mammal genus survives on average only about 5 million years, and individual species a much shorter time. The oldest completely unchanged animal species is probably the crustacean *Triops cancriformis*, a species of crab that lives in ponds all over the world, and whose form has more or less remained unchanged for 180 million years. The reason for its success is that it requires no male: The females have hermaphroditic sex organs and fertilize themselves. In this way they save valuable time. If their puddles dry up too quickly, they can survive such a catastrophe unharmed; the eggs can survive as much as fifteen years of dryness.

Edaphosaurus

In the effort to understand past life, the science of extinct life forms (paleontology) uses a phenomenon that has the advantage of translating the enormous numbers of geologic history into small ones: fossilization. Let us imagine a lagoon in the Jurassic period, whose fine, chalky silt slowly sinks to the bottom and builds up sediment. A primeval bird dies on the lagoon and is embedded in fine mud and preserved without contact with the air. The remains of the animal's body will either be petrified by chemical processes on the pond bottom or a natural cast will be made in the hollows remaining after the dissolution of the body. In this way the stone record books of the earth's history are printed.

The short, digital memory of the present

In comparison with petrified fossils, the durability of electronic data supports is almost laughable. At room temperature (20° C.) and 40% humidity, data can be retrieved for the following lengths of time:

VHS-tape:
15-20 years
Microfilm (normal quality):
20 years
CD-R (recordable):
30 years
CD-ROM (read-only):
50 years
Microfilm (archive quality):
200 years
Newspaper:
20 years
Special long-duration paper:
500 years

Can it be, then, that of our current knowledge nothing will remain? [2]

Every forty years, the developing fossil apparently sinks a millimeter deeper. After 20 million years, 500 meters of stone lie over it. Sooner or later, the geological elevator changes direction, and the long, slow process of sinking down is followed by a rise at the same speed. After 120 million years in the grave, the petrified primeval bird is suddenly found by a worker in a quarry. Then paleontologists work at the find (if they are lucky and are called in), with toothbrushes, small spatulas, and sandblasters, often for weeks or months.

Archeologists and astronomers, climatologists and molecular biologists, zoologists and ecologists all help the paleontologists in their work. The traces of natural events, changes in the landscape and methods of cultivation are preserved in sediments, lakes, and glaciers. The various disciplines decode different fragments of the same text, and work together to fill in the gaps in our knowledge. Every detail is important. Paleontologists even examine fossilized excrements (called coproliths): Are they lumpy, crumbly, or spiral-shaped? The answer to this question may tell us something about what the animal ate for breakfast.

Allosaurus

The Pompeii of primeval times is the Messel pit mine, 25 kilometers south of Frankfurt am Main, Germany. With about 10,000 finds (including seventy primeval horses), it is probably our most important archive of the development of vertebrates, especially mammals, and thus also of humans. It records a period of time from 57 million to 36 million years ago, when the giant reptiles had died out and the animal and plant worlds were undergoing revolutionary changes. The depression was then an inland sea in a tropical forest, and today it is an encyclopedia of the life of a whole epoch. The characteristic Messel oil shale is as soft as praline and contains unusually well-preserved fossils. One hundred species of vertebrates have been found there, including forty species of mammals.

When the layers of slate crack apart, birds, fish, reptiles (including crocodiles), amphibians, insects, and plants are revealed. Often even hair, feathers, and soft tissues, such as that on bats' wings, can be seen in the form of dark outlines on the slate. No doubt much work for paleontologists remains in the deepest strata at Messel: At the current rate of excavation, it would take 40,000 to 50,000 years to examine all the stone, to organize the finds, and to count the various species. In 1995, UNESCO cited the Messel depression as part of the world's cultural heritage. A few years earlier, the officials in charge of the quarry had tried to transform it into a local garbage dump (which would have made it possible to draw conclusions about our contemporary zeitgeist some 50 million years from now).

Our attempt to introduce order into primeval life is full of both errors and enormous gaps. Nonetheless, it offers a limited view of the past. The dinosaurs (about 450 species of which have been described) shared their habitat with a huge number of insects, spiders, and vertebrates. They all lived in a landscape covered primarily with ferns, horsetails, and conifers. The climate was generally warmer than it is today.

But just how many species of dinosaurs were there? And how many individuals could the planet support at the same time? What kind of habitat did the various giant reptiles require? Which species lived in herds, and how large were these herds? In dealing with such questions, science is still groping its way. One problem is that we know very little about the productivity of plants in that era and therefore about the fundamental food supply. We can only speculate: an enormous plant-eater such as the brachiosaurus (which has a long neck and small head) probably used its barrel-like belly as a huge fermenting chamber. This transforming bioreactor might have needed no more plant material than a modern-day elephant, because the fermentation of bacteria produced the proteins and fatty acids on which the huge creatures actually lived. Also, because of the warm climate in which it lived, a brachiosaurus would have used less energy for heat than does, for example, a modern cow. Millions of these giant reptiles were able to feed themselves and to range over the land in huge herds.

Corythosaurus

A playful calculation might show that an inventory could produce some surprises. If we assume that one dinosaur generation lasted, on average, fifty years, and that at a given time there were only a million dinosaurs in the world (for comparison, recall that today there are more than 1.3 billion cows on the planet), then 2.8 trillion individual dinosaurs would have lived during the whole period (there could also have been many more or many fewer). The number of all modern human beings who have lived over the past 50,000 years is estimated to be about 106 billion (see also p. 254). So we can conclude with reasonable certainty that considerably more dinosaurs than humans must have lived on Earth in the course of its history.

The unrealizable dream of all paleontologists is to travel back in time in order to count, weigh, and measure the animals themselves. Then they could verify their intellectual reconstructions of extinct species by comparing them with the living animals. How fast does a plant grow; what climate does it require, was it edible? How did an animal feed itself, how did it move, how fast could it run, and how hard could it bite? One of the most sensational events in twentieth-century zoology was a discovery that made this kind of time travel possible. It occurred in 1938, on the coast of southern Africa. Among the catch of a fishing boat from East London that had tied up in the harbor was found a coelacanth (genus *Latimeria*). Experts knew this animal only from fossils from 150 million to 400 million years old – the coelacanth was supposed to have become extinct some 100 million years ago. Yet this primeval fish had turned up completely unexpectedly, and its basic structure did not differ significantly from that of its fossil predecessors. Scientists caught other coelacanths and observed them. When it was swimming, the fish's lobed fins moved like the legs of four-legged land vertebrates. *Latimeria chalumnae* (the coelacanth's scientific name) confirms hypotheses regarding the way life might have moved from the sea to land.

Diatryma

Fishermen in the Comoros had already known this fish for generations, and had classified it as "not very useful." This messenger from earliest times smells bad, its flesh is very oily, and it has extremely large, rough scales. The latter were the only thing the fishermen could use; apparently they could be employed to rough up punctured bicycle inner tubes.

A fisherman does not consult a zoological classification manual as he empties his nets; neither does he consult the fossil imprint of a fish on a 150-million-year-old piece of stone that was dug up in some quarry thousands of miles away. Conversely, paleontologists and zoologists who are looking for primeval creatures do not normally rummage around in fishermen's toolboxes. As the classic example of the coelacanth shows, we could benefit greatly from a stronger interconnection between knowledge in differing specialties and areas of life.

Brontosaurus

See also plates:

A Changing World Page 24
Humans, a Career
Page 26

Sources for text in margins:

1 New Scientist, 27 July 1998

2 National Media Laboratory, St.Paul: Long-Term Preservation of Digital Materials, 1996

The Great Chroniclers of Nature: Changing Reasons for Studying Nature

For thousands of years, humans have tried to learn about the creatures around them, to order and classify them. The various ways of viewing nature reflect changing worldviews.

Aristotle
384–322 B.C.E.

In 1868, during the annual hunt in the sun-drenched karst landscape near the town of Santillana del Mar in northern Spain, a fox was spotted. One of the dogs suddenly disappeared, and he was heard whining from the depths of a foxhole. Fortunately, the hunters were able to extricate the dog from a subterranean passage that ended in a huge hole. A short time later, on the ceiling of this hole, a twelve-year-old girl discovered splendid paintings of animals in brilliant red, brown, and yellow colors. Her surprised cry, "Papa, mira, toros pintados" ("Papa, look, painted bulls"), became known all over the world. The girl had discovered the first important prehistoric cave paintings. The cave was named "Altamira" in her honor.

The Altamira paintings were made more than 20,000 years ago by hunters who thereby immortalized themselves as both early artists and the first systematic thinkers. In order to survive, human beings in the Old Stone Age had to observe nature, and how carefully they did so we can see in the numerous cave paintings (in northern Spain and southern France alone, 120 caves with paintings have been discovered). The hunters painted chiefly animals that they hoped to put on their menus – aurochs, deer, horses, reindeer, bears, ibexes, mammoths, elephants, rhinoceroses. For the first time, living animals made images of other living creatures. The cave paintings provide us with information about the appearance of of individual species.

Acquiring mastery of depiction was at the same time an intellectual step toward domesticating animals, which began gradually about 10,000 years ago. When they started raising livestock and farming, human beings refined their knowledge of animal and plant species. While they had at first distinguished between dangerous enemies and game that could be hunted, between inedible and edible plants, now they were concerned with animals and plants that were suitable for keeping and breeding. Human beings began to manage both the numbers and the development of animals, ranging from cattle to silkworms. Plants such as wheat and pumpkins were grown and bred. The path toward more precise research, counting, and recording of plants and animals was thereby laid out. In a barter economy, these creatures also served as currency.

Pliny the Elder
23–79 C.E.

For example, a tomb inscription more than 2,000 years old shows that a rich Egyptian had 3,998 cattle. In addition, he possessed 1,135 gazelles and 1,308 sable and addax antelopes, which he had been trying to domesticate. Historically, most inventories have been made with a view to economics, not zoology. These sources – which include early tablets bearing cuneiform script, the large registers kept by feudal lords, merchants, and craftsmen, contracts for the use of agricultural land, and leases – provide precise and historically useful figures. They tell us about hunting and raising livestock and even allow cautious statistical estimates. Very accurate records were also kept by military men: Europeans first saw elephants in a battle near Gaugamela (in modern Iraq). Fifteen military elephants belonging to Darius, the king of Persia, were captured by the Greeks. Later on, as the campaign moved toward India, 300 more were captured.

Economically useful animals grew increasingly valuable. It became more lucrative to keep animals than kill them. The oldest known document on veterinary medicine (2600 B.C.E.) comes from

Egypt. The new cultural values are shown by the fact that humans began to keep animals and plants not only as food sources and means of transportation but also for their peculiar features and beauty. The pharaohs of ancient Egypt, and most of the rulers of that time, including King Solomon, had large collections of animals. A Sumerian-Akkadian dictionary (2000 B.C.E.) and the Assyrian library of King Assurbanipal already listed the names of several hundred animals. In his treatise on hunting, Xenophon described various dogs and hares as well as Keros's zoo and hunting reserves, together with their animals. An earlier predecessor of modern zoos was founded in China around 1100 B.C.E. and called "The Garden of Wisdom." The garden spread over 400 hectares, and in it the emperor kept not only elephants, rhinoceroses, and lions, but also birds, turtles, and fish. The distinction between a collection of curiosities and serious zoological research had not yet been established.

Conrad Gesner
1516 – 1565

In 335 B.C.E., the Greek philosopher Aristotle created the first systematic summary, drawing on the rather random and limited descriptions and insights regarding various animal species that had been made during earlier periods. In his great treatise *Historia animalium*, he described at least 300 vertebrate species so precisely that modern zoologists can identify them (altogether, he mentioned 549 species). In its awareness of constant change in nature, the *Historia animalium* anticipates our current way of seeing living things.

Some 400 years later the Roman writer Pliny the Elder published the thirty-seven volumes of his *Historia naturalis*, in which he made extensive use of Aristotle's work. It contains much solid zoological knowledge, though beautiful mermaids pop up here and there and winged horses gallop over the pages. In the Middle Ages, such fabulous creatures got the upper hand: Natural science took a back seat, yielding the limelight to myths and legends. Illustrated bestiaries entertained credulous readers, and herbals combined precise description with a belief in magic. Wild animals satisfied the public's hunger for sensational sights and were among the favorite official gifts for emperors, kings, and popes.

Conrad Gesner, a Swiss physician and naturalist born in 1516, finally took up where Aristotle and Pliny had left off. Using the works of medieval Arab and contemporary scholars, he composed his *Historia animalium* (1551–1587), and part of a similar volume on plants. The 4,500 pages of his *Historia animalium* provide a systematic inventory of all the species known up to 1550. Each species is thoroughly described in eight sections dealing with its appearance, its behavioral habits, its uses for food and for medicine, etc.

The Renaissance and Galileo's discoveries at the beginning of the seventeenth century fanned the sparks that resulted in the Enlightenment. At the same time, science acquired new tools: The telescope and the microscope changed our way of looking at the cosmos and the diversity of life. Natural scientists began to count plants and animals and to classify, systematize, and describe them. Among the leading figures in this enterprise of inventorying and systematizing were:

Carl von Linné
1707–1778

– Carl von Linné (1707–1778), a Swedish scientist commonly known as Linnaeus. In the middle of the eighteenth century, he laid the foundations of the system we still use for classifying the innumerable organisms on Earth. His system is hierarchically arranged in classes, orders, families, genera, and species. In addition, Linnaeus gave each life-form a Latin name clearly indicating the genus and species to which it belonged. Altogether, Linnaeus named and classified 8,000 different plant species, as well as several thousand animal species, including 828 mussels and mollusks, 2,100 insects, and 4,777 fish, birds, and mammals. His magnum opus, *Systema naturae* (1758) is a milestone in natural history.
– George-Louis Leclerc, Comte de Buffon, a Frenchman and Linnaeus's arch-rival (1707–1788). He was unable to secure the general adoption of his competing classification system (which was constructed more as a network), but his comprehensive, thirty-six-volume work, *Histoire naturelle*, is nonetheless one of the greatest achievements in science. Buffon tried to cover the whole spectrum of nature in the three realms of animals, plants, and minerals. He began at the highest level, and because of the enormous diversity of animals, he never got as far as invertebrates or plants (for this reason, his colleagues published a few complementary volumes after his death). But Buffon's wide-ranging conception of

nature was also his strength. He also discussed the diversity of human beings, their languages, their arts, and their social systems.

– The Frenchman Jean-Henri Fabre (1823–1915). Whereas Buffon taught us to see nature writ large, Fabre taught us to see greatness in the smallest things. Victor Hugo called him "the Homer of insects." The ten volumes of Fabre's main work, *Souvenirs entomologiques*, describe emperor moths, praying mantises, sand wasps, mason bees, and jumping plant lice. His detailed and entertaining inventory makes the reader into a sort of Gulliver, traveling in a strange, multiform, and occasionally eerie world. In his old age, this devoted researcher, with his small stature and dried-out skin, began to look a little like an insect himself. He preferred to work in southern France, under blue skies and listening to the cicadas. When asked whether he believed in God, he answered: "I don't believe in him, I see him in Nature."

Comte de Buffon
1707 – 1788

Fabre's humility is still relevant, though – or precisely because – genome researchers are seeking to decode the complete genotype of life-forms, such as the "weed" *Arabidopsis thaliana* and the fly *Drosophila melanogaster*. Both of these species are ideal for research because of their relatively small genomes. Numerical arrangements – for example, those based on the number of stamens and carpels – were already used by the great natural scientists to distinguish and classify plants. This has led to a downright countdown of life: Using whole batteries of computers, American scientists have decoded the 160 million components of the *Drosophila* genome, and an international team has done the same for the thale cress. However, when the last component is identified and classified, basic questions will still remain: How is conduct related to genes? How does life work? After the great deconstruction, scientists long to connect things again and to explain them, as Aristotle and Darwin did in their times. Human beings see themselves and their fellow creatures not only in their visible forms, but also in relation to their goals and thinking, which are always changing. If it is worth drawing up a balance sheet of life, not only biological reagents count, but also intellectual structures and worldviews. A contemporary example of this is the spread of ecological approaches. Awestruck researchers, such as

Jean-Henri Fabre, helped make the idea of conservation known to a broader public in the nineteenth century. Then in 1910 the Swiss scientist Paul Sarasin suggested a world organization "to protect nature from the North Pole to the South Pole, on continents and in the oceans." After World War II, this idea led to the International Union for the Conservation of Nature (IUCN; today known as the World Conservation Union), which in 1966 published the first list of twenty-six endangered mammal species. Animal and plant censuses, which had previously been conducted for economic reasons, now took on a completely new, ecological orientation. Even old hunting records and whaling statistics began to be studied from a different point of view.

For example, the old inventories made by the Hudson's Bay Company – which was established in the seventeenth century and which kept careful records of the number of fur-bearing animals killed – are valuable for scientific research. Ecologists noticed that in the inventories of snowshoe hares in an enormous region reaching from Alaska to Newfoundland, the numbers rose and fell in ten- to twenty-year cycles, and in some areas these variations reached three-figure magnitudes. The populations of Canadian lynxes, which eat these rabbits, were then found to vary in accord with the same cycles. This makes it easier to distinguish between population changes that are caused by humans and those that have natural causes.

Today, the World Conservation Monitoring Centre (WCMC) collects data from both historical sources and current statistics. The center was founded by the IUCN and the World Wide Fund for Nature (WWF). The goal is to assemble data as precisely as possible in order to provide a basis for action to protect endangered species. In this way we return to the oldest legends: twenty-first-century animal censuses have come back to the theme of Noah's ark.

Jean-Henri Fabre
1823–1915

See also plates:

Small Animals Shape the Earth	**The Ant as a Model of Success**	**Hot Spots of Biodiversity**
Page 20	Page 22	Page 32
		Survival in Zoos Page 38

Human Censuses: Changing Reasons for Studying Human Populations

For over 4,000 years rulers have been counting their subjects. In earlier times, a large population was always considered proof of a flourishing and rising community. Today it is the overall world population that is at issue – and its growth is a matter for concern.

One, two, three, four: Rosencrantz and Macduff keep their eyes glued to the screen. The two rhesus monkeys work at Columbia University. On the monitors in front of them appear four images that represent four objects. The primates' job is to touch first the image that has one object, then the one that has two, and so on. The question is: Can animals count? Rosencrantz and Macduff's performance suggests that our hirsute cousins can pick out the images in the right order with a high degree of accuracy.

While their enthusiasm for this work was initially encouraged by rewarding them with bananas, after a few months they mastered the exercise just for fun – and learned to pick out up to nine objects. According to the psychologists who observed the monkeys playing this video game, we can conclude that there is a natural understanding of counting. Most scientists previously believed that this kind of numerical capability could not be developed without language.

Rosencrantz and Macduff's performance is of interest to anyone trying to trace the development of prehistoric societies. An understanding of counting makes it possible to order and control things, to divide up work and exchange things. We are astonished when we look back on the inductive leap through which a band of primates transformed itself into a civilization of 6 billion people (and Rosencrantz and Macduff should remind us that not only humans count).

It is almost as though humans had taken an express elevator to the top of a skyscraper and left their poor relatives behind on the ground. Humans have about 40,000 genes – and only 1.5 percent of their gene sequences differ from those of chimpanzees. Humans and apes have common ancestors but went their separate ways about 5 million years ago. This only seems to be a long time: If humans were to make a chain of generations, the son taking his mother by the hand and she her father, and so on, after about 250 kilometers we would be shaking hands with a chimpanzee (one generation is here assumed to be twenty years long and each person to be standing one meter apart).

Somewhere in the last ten kilometers of this chain, humans began to speak clearly and to count out loud. What one person says to another has to be treated with care, especially when we are dealing with about 15,000 languages. Nonetheless, what remains partly unsaid is highly informative: Some ancient tribes still have no words for numbers larger than four. Anything more than four is called "many." An archaic calculation aid has remained, despite all progress: the five fingers of the hand.

Primeval clans and bands organized to conduct common hunts and shared their prey – and for both these tasks rudimentary counting was helpful. We can assume that dividing up a dead mammoth between two hungry families was a standard counting job. About 50,000 years ago meticulous contemporaries began to record quantities by cutting notches on sticks. In this way the foundations of systematic bookkeeping were laid. This practice also shows that there was an increasing relationship between counting and private property. Early division of labor, the laying up and managing of stocks, as well as planning for hostile encounters with competitors, necessarily raised the question of how many people were in a clan or a tribe. Putting the best interpretation on this, we can see in it the birth pangs of population censuses.

Anyone who wants to know how many people lived 10,000 or 20,000 years ago has to collect evidence, just like Sherlock Holmes. Specialists in prehistory are able to make even a hand axe or the remains of a campfire speak. Discoveries and excavations of bones, plant remains, tools, courtyards, and villages allow us to make retrospective conclusions regarding the conditions under which humans lived at a given time and to make a rough estimate of their numbers. In this way the population density in an area over which hunter-gatherer communities ranged is estimated to have been from 0.2 to 0.02 humans per square kilometer. We can get some inkling of this by traveling through the Western Sahara. Today, the human population density in this part of Africa is about 1.0 per square kilometer. Although this is significantly higher than the population density in the Stone Age, it still provides quite a lot of elbow room. With the development of agriculture, in many places population densities suddenly became much higher. It is estimated that an early farmstead fed about eight people. In areas where agriculture was established, the population density rose in early times to about two to twenty-five persons per square kilometer, and sometimes considerably more.

New methods of analyzing genetic heritage can now help us reconstruct the development of prehistoric populations. Genome researchers interpret certain messages in our cells as an evolutionary clock – and then let them run backward through the generations. Some microbiologists think they have reached as far back as Eve: According to them, the ancestor of all human beings can be determined on the basis of a characteristic part of the genetic heritage. The first woman is supposed to have lived in Africa more than 100,000 years ago. Humanity would therefore be a large family with a common primeval grandmother.

Adam has not yet been nailed down, but Argentine molecular biologists believe they have determined the primeval father of most Native Americans. They maintain that about 85 percent of all original South Americans and almost half of all native North Americans sprang from a single man. He may have been a member of a small group of earlier pioneers who crossed the Bering Strait – which was then dry – to North America about 22,000 years ago (it is presumed that a second wave of immigration took place about 12,500 years ago). Many experts believe that the continent was previously unsettled; in current terminology, one would say that the population density of North America was zero. Instead, elephants and lions made the Wild West dangerous.

"Molecular archeology" is steadily perfecting its ability to use human skeletons (and the genetic information they contain) to establish family relationships over many generations. This makes it possible to learn about the emergence, migrations, spread, and dying out of whole peoples. Critics even fear that claims to property, dominion, or superiority could be inferred from this kind of investigation, and might be used for political purposes.

Some molecular biologists believe that genetic analyses suggest that during the last Ice Age in the Pleistocene epoch (about 100,000 years ago) the number of our early ancestors might have decreased by 90 percent, to about 10,000. We often forget that for the largest part of our career by far, we were an endangered species. Climatic shifts, hunger, and illnesses constantly threw early humans into existential crises. Catastrophic setbacks and long phases of very slow growth alternated with sudden leaps forward like the one that occurred after the discovery of agriculture.

Science is now able to offer something like a global retrocensus for the early age of agriculture; the world population in the year 8000 B.C.E. is estimated to have been about five million to six million people, less than the population of the Chicago metropolitan area. Driven by the desire for a better life and by curiosity, early humans had already settled all the continents (except Antarctica), with certain centers enjoying special popularity. Jericho was a metropolis in 7800 B.C.E., when the city's population is estimated to have been 2,700.

The first known census was made in China in 2255 B.C.E. The occasion was a natural disaster: A flood had

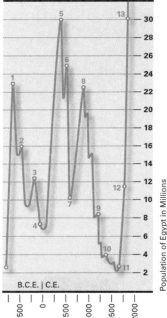

B.C.E. | C.E.

Population of Egypt in Millions

Changing History
In the past, the human population was repeatedly decimated by catastrophic setbacks. This is clearly shown by the example of Egypt:
1 Persian invasion
2 Macedonian invasion
3 Roman invasion
4 75 C.E.
5 Epidemics begin
6 Arab invasion
7 End of the plague
8 Return of the plague
9 Plague spreads
10 Turks conquer Egypt
11 Turnaround in the trend, 1800
12 1907
13 1966

drowned or left homeless a large number of people. The rulers decided to get an idea of the extent of the damage by setting up a registration system. The motive was not selfless concern, of course, but was rather for the purposes of taxation. It was a matter of locating the surviving taxpayers.

The census takers in the earliest high cultures generally limited their efforts to identifying able-bodied men who were suitable for military service or were heads of households. Women and children were generally not counted, though slaves were, since their number indicated the owner's wealth. Fowl, cattle, and the size of the fields owned were carefully recorded by the bookkeepers.

The Egyptians were particularly good at collecting taxes. Every two years they counted the city dwellers and farmers living in the agricultural area along the Nile, as well as how much gold and cattle they possessed. A not insignificant portion of the latter had to be handed over to the pharaoh. Paintings on the walls of tombs show farmers delivering eggs or livestock. They are accompanied by scenes of punishment that make it clear that there were disagreements regarding the amount of the tribute to be paid. People in ancient Babylon, Greece, and Persia are believed to have had to endure similar taxations.

It was hard to escape the census takers in ancient Rome and its provinces, as well. A Roman census conducted in the area of the modern state of Israel became famous through the Christian story of the nativity; according to the biblical account, this census was taken at the time of Jesus' birth. This is the zero point from which the modern calendar begins. In the New Testament, we read, "In those days a decree went out from Caesar Augustus that all the world should be enrolled. This was the first enrollment, when Quirinius was governor of Syria. And all went to be enrolled, each to his own city. And Joseph also went up from Galilee … to be enrolled with Mary, his betrothed" (Luke 2:1–5).

The "silent night, holy night" during which, according to tradition, Jesus was born in Bethlehem was less silent and peaceful than it is often believed to have been. In his work, *De Mortibus Persecutorum*, the Roman writer Lactantius described the census takers' methods: "Everywhere one heard the cries of those who were being tortured and beaten to make them answer the questions; sons were set against fathers, wives against their husbands ... and when they were overcome by the pain, people wrote down taxable possessions that did not exist."

Today, such ancient sources allow us to make a rough estimate of the world's population at that time. According to current views, the opening balance in the year 1 was about 300 million people (or slightly less than the population of modern-day North America). One thousand years later, in the Middle Ages, the world population had probably increased very little. We are indebted to William the Conqueror for a well-documented census taken at that time. In 1086, the Norman invader had England surveyed. As a result, 1.25 million inhabitants were registered, and according to modern estimates, another three-quarters of a million evaded the census. Within the 13,418 localities mentioned, the area covered with forests is estimated at 15 percent, and mills, vineyards, salt houses, and beehives are also listed. Everything is added up to produce a total "wealth of the land" amounting to 73,000 pounds. On this basis a new land register was made, with the help of which 200 Norman barons appropriated the conquered land and its inventory. This record (which may be seen in the Public Records Office in London) was supposed to be valid "until doomsday," which is why the embittered and mocking Englishmen called it "The Domesday Book."

Doomsday did not occur, although during the subsequent centuries illnesses and wars took on a apocalyptic appearance. It was not so much the living as the dead that were counted up. In four years, starting in 1347, the bubonic plague reduced the population of Europe by 40 percent; about 28 million people died. During the Thirty Years' War (1618–1648), up to 80 percent of the population died in areas such as Brandenburg and Pomerania. The depopulation was so dramatic that wolves, which had previously been driven out of these regions, returned and howled in the woods.

Things didn't go much better in the New World – at least for some of its inhabitants. This is shown by the first American census of slaves. In March 1619, "fifteen men and seventeen women in the service of planters" were counted in Virginia. From then until 1808 (when it became illegal to import slaves into the United States), 500,000 Africans followed them. The slave trade constitutes a whole chapter of the history of censuses (beginning with the Persians and Egyptians).

Between the sixteenth and the eighteenth centuries, mercantilism led in many places to an economic upswing and an altered world-view. Governments abolished internal tariffs, standardized weights and measures, and promoted exports. At the same time, the citizens of a country came to be seen as national capital, which should be increased as rapidly as possible. This brought about a regular boom in new censuses, which changed in character because they were oriented toward the future and contained an element of prognosis. To better judge a population's potential for growth, the census takers started recording not only the number of persons but also data on age, sex, family status, and profession.

The first census of this new kind was conducted in 1666 by the government of French Canada. It was followed by censuses in Iceland (1703), Sweden (1749), Denmark and Norway (1769), the United States (1790), and England and France (1801). Germany began to take regular censuses in 1875, and Czarist Russia made its first attempt at a census in 1897. To the delight of proponents of mercantilism, between 1800 and 1900, the fastest growing societies were in Europe and North America. Europe doubled its population, and America increased its population twelvefold (thanks to large-scale immigration). The initial population recorded by the United States Census Bureau in 1790 was 3.9 million inhabitants (more precisely, 3,929,214). Each household had to provide the same information: "Family name, number of persons in the household, free white men over sixteen years of age, free white women, slaves, and other persons."

In the twentieth century, the nature of censuses changed again because great ideologies took control of this instrument. In totalitarian regimes, counting and control were closely connected and became a tool of the police state. In the censuses conducted by the Nazis, it was not just a matter of counting people but also of transforming them in accord with the ideas of their new masters. Thus the number of children "biologically valuable women" were supposed to produce was calculated, and the statistical foundation for the claims regarding "people without space" and for World War II were prepared. Questions concerning "racial membership" and "relations with Jews through marriage" were later used as a basis for ethnic selection and for the annihilation of the Jewish population and other minorities.

The 1937 census in the Soviet Union also is notorious. After production statistics and planning figures became fetishes, population data were ultimately used to demonstrate the superiority of the Soviet system over that of its capitalist competitors. Socialism's motto was the same as that of mercantilism: The more people born, the better. Reports of high birthrates and the rapid growth of cities were seen as an index of progress and increasing prosperity. The dark side of Soviet rule, such as poor living conditions, low levels of education, and high rates of infant mortality, was kept hidden from view. To the dismay of those in power, the census also showed that the food shortages of 1932 and 1933 were caused by the state and cost the lives of 5 million to 7 million people. Stalin's henchmen summarily punished the messengers who brought the news: The organizers of the census – many of whom were people with decades of professional experience – were arrested and shot.

Today, censuses have developed into a science – demography. It works with complex mathematical formulas and statistics, cleverly devised questionnaires, and computers. Its fundamental task has also undergone a major paradigm shift; the belief that a growing population means progress for a country has been deeply shaken. Instead, family planning and voluntary limitation of the number of children produced are visibly becoming the chief focus. Merely counting the population is no longer nearly enough; rather, demographers seek to give political decision-makers crucial help by providing three-dimensional descriptions.

For example, the scientists at the Population Reference Bureau (PRB) in Washington, D.C., and at the World Population Foundation (Stiftung Weltbevölkerung) in Germany have developed a prototypical profile of a young African woman and a young European woman. On the basis of average demographic values, they can describe a typical life in each place. Take Eden, who lives in the village of Moulo, in Ethiopia. She has five sisters. She has her first child at the age of sixteen, and leaves school. At eighteen, she marries, and her second child dies five months after it is born. She works at home. At twenty-eight, she has her fifth child, who dies at birth. At thirty-three, she becomes a grandmother; she has had eight children herself, six of whom are still alive. At fifty-two, Eden dies. Julia lives in Kassel, Germany. She has one sister. At sixteen, Julia begins her first relationship, and starts taking birth control pills. She studies at the university and marries at the age of twenty-six. At twenty-nine, she has her first child, and three years later a second child. She becomes a grandmother at fifty-eight, and dies at the age of seventy-nine.

The corresponding societies can be described just as graphically. For example, young and rapidly growing societies like the one Eden lives in are pyramidal, with a broad base consisting of young people. In contrast, the populations of older, more stagnating countries like the one Julia lives in are more columnar; the number of young people at the bottom corresponds to that of the elderly people at the top.

Pyramidal and columnar societies have completely different demographic equilibria. As a result, they have completely different problems: Young countries are scarcely able to provide for and educate the increasing numbers of people who live in them; stagnating nations suffer from the fact that the decreasing number of young people have to take care of a steadily increasing number of old people.

The distribution of public funds, the rights of minorities, the construction of retirement homes or hospitals, urban planning and highway planning, the training of teachers – all of these depend on reliable demographic data. Companies also increasingly make use of

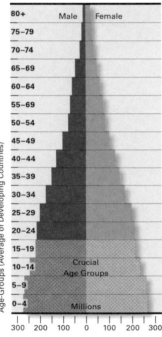

The shape of a "young" population
The population of quickly growing developing countries is pyramidal, with a broad base of young people. Many of these countries face great problems in providing proper care for this rapidly reproducing population. 2

demographic information in developing their business plans. The present chapter in the development of human beings describes the creation of consumers – and in the long run, their numbers and living conditions determine stock prices.

In the United States, India, Kenya, and Switzerland, a census is taken every ten years. The relevant laws (and, in the United States, the Constitution itself) require this. In Japan, Turkey, Canada, Australia, New Zealand, and South Africa, the government usually counts the citizens every five years, and in France about every seven years.

Neither money nor labor is spared in making these censuses. The United States Census Bureau currently estimates the cost of a single census to be $4 billion to $6 billion. To count their citizens, the Chinese require an army of 25 million helpers (if they ever refuse to do this work, the Chinese government would have to hire the whole population of Venezuela to replace them).

The difficulty of conducting a precise census can be shown by a comparison with a football stadium in which at halftime the precise number of spectators is supposed to be announced. The number of tickets sold cannot be used for this purpose, because some people have climbed over the fence or used counterfeit tickets. Conversely, persons who have bought tickets beforehand may not have come to the stadium. Counting the people in the stands will not do the trick, either, because they wander around or go to buy hot dogs. And in a country people move about at least as quickly as in a football stadium. For this reason, in 1947, the government of Iraq had the following idea: It locked up the whole population for the duration of the census. It closed the country's borders and imposed a curfew from morning to evening.

Because demographers do not want to make themselves unwelcome, they have to accept a certain amount of imprecision. For example, the 1990 United States census is said not to have included 8.4 million citizens

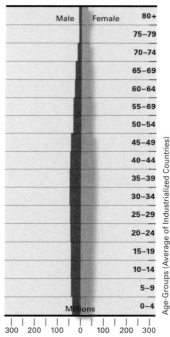

The shape of an "old" population
For industrialized countries that are growing slowly or not at all, the population is columnar, because the number of older people is much larger and built upon a narrow base of younger people. The rising cost of providing health care and care for the aged is causing problems for these countries. [3]

and to have counted 4.4 million twice. This can be inferred from random samples and projections. A large part of these overlooked Americans were under eighteen years of age and so belonged to a group especially important for the future. Although in the United States the law calls for a strictly individual counting of citizens (which is difficult to achieve), demographers consider a combination of actual counting and projections from random samples to be more accurate. Even so, in their calculations they assume a tendency to undercount which can amount to as much as 5 percent of the population.

Now many nations other than the powerful United States of America are counting their citizens again, including one of the most populous (India) and one of the richest (Switzerland). Whereas 150 years ago only about one person out of five was actually recorded, today there are hardly any blanks left on the demographers' maps. To make it possible to estimate the population of remote regions, satellites may be used that can provide information about communities that do not even appear on maps.

All the available international data will be evaluated by the population department of the United Nations, with the help of scientific institutes throughout the world. Whereas the results of national censuses were previously the chief focus, today the overall world population figure is seen as the key point in public discussions of the future of humanity. This figure, together with the human capacity for reasoning and innovation, will determine how successful we can be in providing food and care for the world population without harming our diverse fellow creatures.

At present, more than 6 billion people live on Earth. About 4.5 billion were added between 1900 and 2000. The world population's clock moves forward at the rate of almost three persons per second. Today, this ticking clock and the photograph from space, in which Earth appears as a small blue island, symbolize our view of the world. From the latter springs a central human project at the beginning of the third millennium: slowing human population growth and at the same time treating natural resources more respectfully. Demographers and astronauts are becoming – without intending to be so – some of the most powerful ambassadors for the understanding of population. We are one world, and without the diversity of our fellow creatures our world would be less worth living in. Even in the global village this discovery can still be counted on five fingers.

See also plates:

Humans, a Career
Page 26
Humans as Victims and Prey
Page 30

Sources for text in margins:

1 J. E. Cohen: How Many People Can the Earth Support?, New York, 1995

2 United Nations Population Department, 1998 / Deutsche Stiftung Weltbevölkerung, 1999
3 Ibid.

Expedition to Planet Earth: The World We Have Yet to Discover

Has everything already been discovered? Actually, more plant and animal species are unknown than are known. Particularly in the oceans, in the canopies of the rain forests, and in the realm of microbes, there are still blank spots on the map of life.

The pressure has risen to 1,013 atmospheres! That kind of pressure could almost collapse the walls. But these walls are made of super-hard steel. Unperturbed, "Kaiko" glides on and dives into the light-less expanse, a world hostile to humans. The Japanese vehicle is un-manned, steered from a distant control center. There, the engineers are preparing for a maneuver they have longed to make for years. Slowly, Kaiko approaches the surface of an unknown planet. Only a few inches to go. Now! Landing successful!

Breathlessly, the observers at the monitors follow what happens next: Their remote-controlled emissary is immediately attacked. It is surrounded by monsters that couldn't be dreamed up, even by a maker of science fiction films. Out of their heads, lanterns grow that throw enigmatic, shimmering greenish clouds of light on the dark-ness. Goggle eyes roll in some of the faces, and below them yawn mouths with long teeth. Half-transparent creatures with sticky ten-tacles undulate before the camera lens, cutting strange capers.

Kaiko is five meters long and weighs eight tons; it is a robot submarine belonging to the Japanese Oceanographic Institute,

JAMSTEC. In many respects, its missions resemble deep-sea diving and space travel. In both cases, humans enter spheres in which they could not survive for a second without important technological help. In each case it is a matter of exploring new worlds. The unknown planet that Kaiko is exploring is called Earth. It should really be called the water planet, since oceans and seas cover 71 percent of its surface. Above these spheres, and especially in areas deeper than a thousand meters, we sometimes know less than Columbus knew about India. When Kaiko is gliding through the water like a ghost ship, the exotic beings of the sea come to meet it: anglerfish, which attract their prey with luminous "fishing rods" on their heads; ribbed jellyfish that search for food with sticky tentacles. Kaiko's cameras pay no attention to their terrifying beauty. Its task is to collect samples of mud from the sea bottom, often more than 10,000 meters below the surface. In the samples from the Mariana Trench, JAMSTEC researchers have isolated many species of microbes – most of which had never been scientifically described before.

Here we find a crucial difference from space travel: Whereas astronauts have so far found no trace of life outside Earth, since the 1960s aquanauts have made a series of spectacular new discoveries. For instance, the sea beneath the layer where sunlight still penetrates is not, as was long believed, an area hostile to life. On the contrary, it is swarming with life, all the way down to the sea bottom. Over the past few decades, daring divers have gone deeper and deeper, in order to reveal the secrets of life in the oceans. Their scientific dreams were inspired by Jacques Piccard's experience in 1960. In the French-built bathyscaph "Trieste," he succeeded in descending to the bottom of the Mariana Trench, more than 12,000 meters below the surface – the lowest point on Earth so far reached by humans. "When we landed, we had the extraordinary luck to see a fish in the beam of our spotlight," he reported. At that moment, a question that oceanographers had long debated was answered: Even higher, developed organisms can exist down there, despite the eternal darkness and enormous water pressure.

Spring 1977, off the Galapagos Islands, at a depth of 2,500 meters: American marine biologists, fascinated by geologists' reports that they had seen photos of strange objects – big, white, and long – undertook a submarine expedition. "Alvin," their submarine, approached a group of hydrothermal vents – crevices and chimneys in the earth's crust through which hot water rises. The scientists found that these geysers were surrounded by a paradisiacal garden. In it lived species of mollusks and crabs never seen before, multicolored sea anemones, and pink fish. And the long, white spectral beings turned out to be huge beardworms that were a meter and a half long, which employed a previously unknown survival strategy. They had no functioning stomach-intestinal tract, but instead met their nutritional needs with the help of countless bacteria inside their bodies. The microbes transform sulfur from the geysers into energy. Unlike plants, which get energy from sunlight through photosynthesis, these worms had found in the darkness a second source of life: chemosynthesis. Once again the textbooks had to be revised.

The oceans, which constitute 99 percent of the volume of the earth's biosphere, continue to be the biggest blank spots on the map of life. Only 1.5 percent of the deep-sea area has been investigated, and the area in which life has been inventoried is smaller still. New animal and plant species are constantly being discovered. Off New England, where undersea explorations are regularly made, biologists have found 798 species of invertebrates in an area no larger than two parking lots. American marine biologist Frederik Grassle estimates that there are more than 10 million species of invertebrates in the deep-sea realm alone. If he is correct, then only about two out of every hundred maritime life forms are currently known. The other ninety-eight live incognito on the blue-water planet.

We are particularly ignorant regarding the microorganisms in the seas. Around the geysers of the Galapagos, bacteria have been discovered that can withstand temperatures as high as 350 degrees Celsius. Previously, it had been assumed that at such temperatures endogenous enzymes would cease to function. Thus the biologists' discovery also had economic potential: Enzymes that could withstand that kind of heat are very useful in biotechnology because they can be used to improve production methods (in the making of drugs, for instance).

The golden bamboo lemur, a prosimian that lives in Madagascar, is up to 80 centimeters long, and weighs as much as 1.5 kilograms. It was first discovered in 1897.

Even the surface of the ocean abounds with surprises. What looks to us like clear water is in fact a chemical, bacterial soup full of life. One liter of it contains about 10 billion viruses, about 5 million one-celled organisms, and about 1 million algae. It is estimated that a third of the earth's total stock of carbon dioxide is stored in the oceans. Yet no one knows precisely which microbe species are involved in such processes. For this reason many more expeditions like those undertaken by Kaiko and Alvin are desirable.

Funds for such expeditions are generally lacking. Even more amazing than the extent of our ignorance of the other inhabitants of the planet Earth is our general indifference to it. The explanation for this indifference may be that our contemporaries vastly overestimate the extent of their own knowledge. After the Enlightenment, there was a feverish interest in new knowledge. New natural laws were discovered, new continents opened up. The intellectual horizon expanded, along with the geographical horizon. But in the twentieth century, which was characterized by globalization and networking, the feeling that everything about the earth was already known became widespread. The age of great expeditions is past, we were told; everything has been discovered, measured, mapped, listed. If we want to find something new, we have to leave the planet Earth and go into orbit. Costly space programs were begun for this purpose. One of the most expensive of these is the international space station. The official estimate of its eventual cost is $40 billion, but skeptics think it will cost closer to $100 billion. For comparison: The United States National Science Foundation has $3 million to support biodiversity research. What is especially deplorable, according to the American scientist Sylvia Earle, is the lack of interest in marine biology. "The oceans and the future of humankind are intimately connected. Ultimately, it is the oceans that determine the world climate and the nutrient cycle." In view of this fact, she considers the budget for marine research to be inappropriately limited. The National Oceanic and Atmospheric Administration's deep-sea program has to get along with only $16 million a year: "That wouldn't be enough to pay for the toilet in an American space shuttle."

So it is no wonder that, so far, very little is known about biodiversity on Earth, whether in water, on land, or in the air. The gaps in our knowledge are so large that scientists can't even be sure about what they already know. How many species of animals, plants, fungi, and microbes have already been described in scientific literature, and thus more or less officially exist? Because there is no international register, it is not easy to determine the number. Today, it is estimated to be 1.75 million. A still more heated dispute concerns how many unknown species might exist. Estimates vary between 3 million and 30 million; Edward O. Wilson, the leading American expert on biodiversity, thinks there may be as many as 100 million. The scientific community generally accepts a purely theoretical estimate of 15 million. If this is correct, then today about nine-tenths of all species are unknown.

The biggest blank spots are located in regions that humans still find difficult to reach, even in the age of global mobility: the deep sea, the canopy of the rain forests, and, the realm of our smallest fellow creatures, microorganisms. Moreover, scientists show far more interest in some of these areas than in others. For example, taxonomists – experts who specialize in the definition of species – fall into two main groups. Most are chiefly interested in large animals with either fur or feathers, whose appearance is appealing. In professional lingo, the latter are called "charismatic megafauna." The other, much smaller group works on bacteria, fungi, mites, protozoa, and nematodes – the inconspicuous life-forms that are nonetheless the key players in the natural economy. In reality, it is the small creatures that rule the world.

It happens that today we know less about biological life and its wealth of variants than Columbus knew about the roundness of the earth. At the beginning of the twenty-first century we can still say that a whole continent remains to be discovered, a continent called biodiversity. Even in densely settled and well-known Europe, there are still treasures to be dug up. Only a few years ago biologists found a new species of frog in a Spanish mountain stream. The description of this new species, *Rana pyrenaica*, made a considerable sensation. Up to that time, it had been assumed that the inventory

The black-and-white tree kangaroo has long been known to the inhabitants of Irian Jaya (New Guinea). Zoologists first discovered it in 1995.

90

of species in temperate climates was virtually complete. Not the least important reason for this is that most of the 30,000 taxonomists worldwide, as well as biologists who identify new species, work in Europe and North America. Only 6 percent work in tropical countries – a flagrant disproportion, given the fact that the tropics contain the greatest biological riches on Earth. In these countries, there is sometimes not a single knowledgeable expert to study less conspicuous, less attractive species, such as invertebrates. The hot spots of biodiversity are in the Southern Hemisphere, while the summits of biological science are in the North, and this division has consequences. An additional problem is that the guild of taxonomists is finding it difficult to recruit enough new members. Fewer and fewer students are interested in specializing in biological systematics. As a result, currently two-thirds of specialists in this field are over the age of forty-five, and this leads people to say jokingly that if a similar aging were found among a rare species, a repopulation program would have to be started.

Are taxonomists an endangered species? An international group of biologists is trying to find a solution to this problem through the Systematics Agenda 2000 project. It recommends the herculean task of inventorying the whole biodiversity of Earth within the next twenty-five years. However, the precondition for this enterprise is that governments invest about $4 billion a year in this large-scale project. In addition, more specialists in systematics need to be trained. According to Professor Horst Schminke, the president of the Deutsche Gesellschaft für Biologische Systematik, 52,000 specialists in this field will be needed to carry out the task for the Systematics Agenda 2000 project. "On a worldwide scale, this should not be an impossible undertaking."

Despite the problems, Schminke senses renewed interest among his colleagues. The chief stimulus was the world summit meeting held in Rio de Janeiro in 1992. In the Convention on Biological Diversity (CBD), the participating governments explained the importance of an inventory of biological life as a precondition for the

Blank Spots on the Map
The following regions have not yet been investigated adequately, even though they probably contain a great diversity of life-forms:
- The Annam highlands (on the border between Vietnam and Laos)
- Chocó (western Colombia)
- Kimberley highlands (northwest Australia)
- Central Amazon region (Brazil)
- Southern slope-humid zone of Papua New Guinea
- Congo basin forests (central Africa)
- Northern and eastern Angola (Africa)
- Forested hill country on the borders between Laos, Myanmar, and Yunnan (Southeast Asia)[1]

latter's preservation. Tangible successes gave taxonomists further impetus. Horst Schminke can cite a series of cases in which his discipline is of practical use or can even be evaluated in terms of cold cash. In California, a species of insect, the dwarf cicada (*Circulifer tenellus*) attacked the sugar beet crop and infected the plants with a viral disease. Because it was wrongly classified, this pest was thought to have come from South America. However, it had no natural enemies in South America. Only when a taxonomist discovered that it actually originated in the Mediterranean area did it become possible to control the dwarf cicada. Another example: Taxonomists are constantly finding wild relatives of domesticated plants. A new variety of wild tomato found during an expedition in the Andes was crossed with a domestic variety, with the result that the quality of the fruit was improved – and this meant additional profits of several million dollars for farmers. In many cases, crossing domestic plants with wild ones increases harvests. In order to identify candidates for this kind of crossing, basic taxonomic research is necessary. In this way, a project like the Systematics Agenda 2000 can help solve the problem of providing food for a growing world population. "Taxonomists' knowledge is in demand again," Schminke says.

These discoveries show that the mission to planet Earth, as envisioned by biodiversity researchers, can be at least as exciting as a flight to Mars. Harvard professor Edward O. Wilson argues that the whole biosphere must be made accessible, in part because it contains many undiscovered treasures that could help treat diseases. About 40 percent of the drugs sold by pharmacists in the United States contain substances derives from plants, animals, or microorganisms. "Yet these substances represent a tiny fraction of those that could be made available," Wilson says. We can use only what we know about. In making the worldwide inventory, researchers nonetheless found themselves in "a race against time," as Michael Balick, director of the New York Botanical Garden, put it. Botanists are competing with bulldozers, and the latter are often quicker than any expedition can be. In 1972, zoologists found a small, inconspicuous frog in small brooks near the Australian city

The Vu-Quang bovine is one of three previously unknown large mammals discovered in the 1990s in Vietnam.

of Brisbane. Further examination showed that the female behaved in an extraordinary way: She carried her young around in her stomach. She obviously had the ability to temporarily suspend production of hydrochloric acid in order to provide this unusual breeding ground. Physicians treating stomach ulcers rapidly formed a team to make this phenomenon useful for human patients. But they were not able to complete their studies; in the meantime, *Rheobatrachus silus* had become extinct.

Again and again, sensational discoveries of new species prove how necessary a research offensive is. The smallest frog in the world was caught recently in Cuba. The fact that more than 700 new insect species are described every year should not be surprising, since several million small species and life-forms probably exist unbeknownst to us. However, great discoveries are constantly being made:

– In New Guinea, a marsupial weighing fifteen pounds trustfully revealed itself to a team of zoologists. Upon closer examination, it turned out to have a new tree kangaroo in its pouch. The natives had known this creature for a long time. They called it Bondegezu, "man of the mountain forests."

– In the Vu Quang Nature Reserve in Vietnam, three new large mammals were discovered. One was a cow; the other two were species of deer. Evidence that one of these existed had long been available: Its antlers hung in the huts of the surrounding villages. But only in the mid-1990s were they identified as belonging to a new species, the giant muntjac. Hunters report seeing these animals quite often.

– Even among primates, new species are constantly being discovered. In the 1980s, three new species of lemur were found in Madagascar, which is not exactly unknown from a biological point of view. These species live only in Madagascar.

– In Brazil alone, seven new monkeys have been recorded since 1990, one of them not far from the city of Manaus in the Amazon area. "If we find new primates so near human settlements," said Russell Mittermeier, president of Conservation International, "we can easily imagine what is still hidden in inaccessible areas."

93

At first glance, it seems entirely impossible to search for specific unknown life-forms in the jungle; looking for a new insect species in the rain forest would be like looking for a needle in a haystack. Wrong! The efforts of insect experts like Terry Erwin, who goes into the Ecuadorian rain forest early in the morning in search of new species, are rewarded. Yet Erwin, who is nearing sixty, begins with a death squad. With the help of a few students, he starts up two machines that spray a cloud of pyrethin, a powerful insecticide, into the air. "Highly effective, but biodegradable," he emphasizes. While the lethal cloud is still floating up the trunk of a giant fig tree, through the vines, ferns, and tillandsias that cling to its branches, dead beetles start raining down into the nets spread out below, where the researchers can examine them. "We need this kind of complete inventory," Erwin says, "as an index for use in maintaining biodiversity in the rain forest."

Erwin's students carefully collect each insect and take their harvest to the Tiputini Biodiversity Station. There the insects are dried and put on pins; then they are examined with magnifying glasses and microscopes. Razor-thin differences in color, the feelers, and the shape of the wings indicate which species each insect belongs to. Entomologists are always fascinated by the way tropical regions keep producing new variations on the same basic structure. And not the least element of their fascination is the thrill they feel when they sometimes examine an insect and say to themselves, "I am the first scientist who has held this in my hands!" In order for researchers to be sure that it is a new species, a few samples are sent to various museums of natural science, where they are definitively classified – a voyage out of the anonymity of the forest canopy into a modest prominence in professional journals.

Erwin's record harvest in twenty-five years of research in Panama's tropical rain forest: Almost 1,200 different beetle species fell from a single tree belonging to the *Luehea seemannii* species (related to the linden tree). More than half of these beetles were previously unknown. Erwin began to extrapolate this relationship. Since beetles are highly specialized and most of them live on only one kind of tree, he multiplied the number of insects by the number of trees and

Custodians of Nature's order
Because of their professional competence and the size of their collections, the museums listed in the table play an important role in the discovery and definition of new plant and animal species.

Major Research Museums and Collections

Museum	Collection
Musée National d'Histoire Naturelle, Paris	68,000,000
Natural History Museum, London	65,000,000
American Museum of Natural History, New York	30,000,000
Naturmuseum Senckenberg, Frankfurt am Main	25,000,000
Museum für Naturkunde, Berlin	25,000,000
Bishop Museum, Honolulu	23,000,000
Field Museum of Natural History, Chicago	20,000,000
Zoologische Staatssammlung, Munich	20,000,000
Zoological Museum, Copenhagen	10,000,000

in this way arrived at an estimate of the total number of different insect species living in the tropics: 30 million. Then he calculated the total number of all animal and plant species living on Earth: The unimaginable result was 100 million! Although quite a few of his colleagues have disputed this estimate (most biologists think it is a far lower, but still impressive, eight-figure number), Erwin defends his view, and even suggests that it may be an underestimate.

In any event, it is no accident that researchers make so many discoveries when they turn to a part of the rain forest that for a long time remained terra incognita – the canopy. There, seventy meters from the forest floor, nature pulls out all the stops. In the canopy, protected like a medieval fortress from its enemies, one finds a fascinating world of ferns and bromeliads, birds and frogs, beetles and snakes. Many of these animals never touch the ground; they live in a genuine enclave – which makes it difficult for researchers to enter. In order to reach it, scientists have to storm the treetops. For this purpose they have tested a series of technical systems: Hanging bridges, such as one finds in the Gunung Kinabalu National Park in Borneo; a sort of inflatable raft, three hundred square meters in size, which is softly deposited on the treetops by a dirigible, as in the rain forest in Gabon; or a complete crane, transported into the middle of the primeval forest, as in Venezuela, can be used. Every means of reaching the crown of creation seems to have been tried.

Swiss helicopter pilots flew the parts of the crane to southern Venezuela, to the Humboldt Research Station near the Indian village La Esmeralda. There work scientists from the most diverse disciplines: climatologists, botanists, ornithologists, entomologists, herpetologists, specialists in bio-acoustics. Together, they use the crane as an elevator to reach the airy heights. In its gondola, they use instruments for making meteorological measurements, tap flower nectar, band birds, map trees, describe mating behavior, and type Latin names into their laptops. Two hundred years after Alexander von Humboldt arrived in Venezuela, high tech makes its entry into the fellowship of discovery. However, the pioneer spirit still survives. Not everyone is able to participate in the general inventory of the swaying forest canopy. Researchers have to be bold, meticulous, and dedicated; their work requires the characteristics of Tarzan combined with those of a stamp collector.

Taken altogether, the various parts of this interdisciplinary project are supposed to provide a more precise mosaic of the tropical superorganism, its ecological interactions, the various forces at work within it, the circulation of food and water. However, like all research results, these raise new questions: Does each insect actually have its own ecological niche, or do several species share a habitat? How long does it take this forest, this home with thousands of tiny rooms, to produce new inhabitants? Robert May sees a discrepancy between the violent human attacks on the rain forest and our knowledge of the latter's natural economy: "The situation clearly calls for a better understanding of the changes that are occurring in the ecosystem as a result of species dying out and the advancement of other species. It is absolutely necessary that we trace the fundamental connection between diversity and stability." If we want to preserve nature, we have to study it.

Three centuries after the Dutchman Antonie van Leeuwenhoek first studied bacteria with the help of a microscope he had constructed himself, no more than 1 percent of all bacteria have been described. This is only slightly better than none, given their importance. Bacteria were the first living creatures on this planet; they discovered photosynthesis and laid the foundations for the existence of animals and human beings: They emitted oxygen into the atmosphere. "Without bacteria," says Bo Barker Jørgensen of the Max Planck Institute for Marine Microbiology, "life in the oceans would to a large extent die out within a year." He points to the metabolic processes in the upper levels of the sea bottom, which operate intensively much as do those in an area of biodiversity, such as the rain forest.

In 1999 a new microbe was discovered in the waters of southwest Africa whose size created a sensation. The "Namibian sulfur pearl" (*Thiomargarita namibiensis*) is approximately as large as the period at the end of this sentence. For comparison, if a sulfur pearl were the size of a blue whale, an average-sized bacterium would be no larger than a newborn mouse. For this reason, it is exceptional in the microworld of bacteria. It is simply their minuteness that has allowed bacteria so successfully to escape the attention of researchers. With the naked eye no one can tell what a particular microbe is and how it lives. Microbiologists have to use petri dishes to grow cultures. But since microbes much prefer to live in their own habitats, many species do not thrive in the laboratory – or always the same species. New techniques are supposed to help make life in the laboratory more pleasant for bacteria, in order to make it easier to do research on them. Using the most refined microsensors, twenty times thinner than a human hair, scientists palpate bacteria's natural environment. In this way they can derive, for example, information regarding the oxygen or hydrogen sulfide content that enters into the creation of artificial environments.

Molecular biologists are increasingly using a new technological method for genetic research. Strands with genetic material are removed from microbes, cut into segments, and the genetic components "sequenced." The resulting sequence is compared with sequences from known species, and the degree of coincidence indicates the degree of relationship. In this way, bacteria also acquire a family tree.

The genetics laboratory, the deep sea, and the tropical forest are only three examples of all the scientific expeditions undertaken to search for the unknown on planet Earth. Scientists have embarked upon an adventure – the adventure of thought, in which the claim, "we already know," is rejected. However, the real test to which they are subjected is that of patience. Taxonomists need a great deal of patience as they collect, examine, sort. As they describe their finds, compare types, classify species. As they doubt, reject, reclassify. "Wisdom begins with calling things by their right names," says a Chinese proverb. For plants and animals, museums of natural science take on this task as wise godparents. Their collections are the largest archive of life. In them are kept – dried, pressed, stuffed, pre-

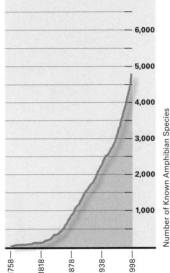

More and more is being discovered
New species are constantly being found, not only in the realm of insects, but also in that of vertebrates. Toward the end of the twentieth century, more unknown amphibians were discovered than ever before. The curve shows the increase in known species of frogs, newts, and salamanders since these species were defined by Linnaeus in 1758. As a result of these new discoveries, the number of known amphibians is now greater than that of known mammals. [3]

served with formaldehyde or in alcohol – the standards for comparison that make correct classification possible. Worldwide, such museums hold around 2 billion examples – 2 billion proofs of nature's enormous creative potential. And each has its own death certificate.

However, natural history museums act not only as cemetery guardians but also as modern research institutes. They use DNA-analysis apparatuses, spectrometers, and high-power microscopes – that is, they function as forensic laboratories in the worldwide search for new species. But while collections strive to modernize themselves by using high-tech methods, the taxonomist's work still relies on the human hand and eye. Consider the daily life of a lepidopterologist, an expert on butterflies and moths, who works in the Natural History Museum in London. Here, taxonomists have access to an impressive collection containing 65 million specimens: meteorites, skeletons, skins and feathers, wasps in cardboard boxes, centipedes in alcohol, petrified fossils in crates and on shelves, stuffed birds in display cases. For a lepidopterologist, the most interesting parts of the collection are the hundreds of drawers each containing hundreds of butterflies and moths mounted on pins, as if frozen in midflight. It costs a great deal of money to establish and maintain such comprehensive collections. The museums' commitments are all the more admirable in view of the fact that they carry out their functions as inventories of species without calling them to public attention. Their biological bookkeeping is not very spectacular; it is not, as journalists would say, "newsworthy." If our lepidopterologist receives a new shipment of moths and butterflies, these jam-packed drawers are his most important tool. He begins his meticulous examination of the newly arrived lepidopters by comparing variations in color, studying the different ways in which the body is shaped; he reads up in the professional literature, lists, and catalogs. A taxonomist is very lucky when he or she recognizes that he is looking at a "type" – that is, a specimen on the basis of which a species is described for the first time, virtually the standard model of a species. It is identified by a colored dot and is considered one of the most precious items in any collection.

However, this classification is still open to doubt. Ultimately, the history of taxonomy is also a history of errors. And so, 20,000

African moth and butterfly species were named on the basis of the collection in the Natural History Museum in London – although probably no more than 3,000 live in the wild. Sometimes it seems as though nature, which excels in camouflage and deception, sets out to fool the experts through imitation and variation. One species of lepidopter, for instance, produces females of four different colors that imitate four different species. Even the taxonomists were taken in by this one, and issued an official list of false species names.

In order to avoid this kind of mistake, taxonomy needs not only modern methods but also a wealth of ancient wisdom when classifying nature. But what will happen when there are no longer any heirs to receive this heritage of experience? The initiators of the Systematics Agenda 2000 urge the training of more taxonomists in universities, in order to reduce the number of misclassifications in the future. Should their proposal be adopted that all the species on Earth be discovered and classified within the next twenty-five years, the need for specialists in this field will increase greatly. Biologists reject claims that this project would cost too much. Harvard primate researcher Irven De Vore argues, "Let us suppose that our global project actually finds intelligent beings – beings whose genetic code is more than ninety-eight percent identical with our own. How much money would we be willing to spend to do research on them?"

As De Vore knows, these beings have already been discovered. Not in space, but on the mission to planet Earth. We call them chimpanzees, orangutans, and gorillas. However, we can only speculate about the size of the primate family as a whole. We don't know how many of our close relatives are still among the anonymous species.

See also plates:

Small Animals Shape the Earth   **The Green Pharmacy**
Page 20   Page 156
Hot Spots of Biodiversity
Page 32

Sources for text in margins:

1 WCMC
2 Ibid.
3 F. Glaw, J. Köhler in Herpetological Review 29 (1), 1998

Extinctions: Losing Species before Their Roles Are Understood

For centuries, humans plundered nature. The result was that many animal and plant species perished forever. In many regions of the earth, this kind of overexploitation is still an everyday occurrence at the beginning of the twenty-first century.

"Relatively few of the countless thousands of crushed, beaten, fallen birds could be picked up, yet wagonloads moved in an almost unbroken chain out of the nesting places, while the ground was still covered with living, dead, and decomposing birds. ... This horrible business went on day and night. ... Goblins in human form, wearing old, shabby clothing, their heads wrapped in crude linen and old shoes or rubber boots on their feet, went about with sticks and clubs and knocked down the birds' nests, while others cut down the trees and broke off the overladen boughs in order to collect the young birds."

What Etta Wilson saw as a young woman in Michigan in 1870 was the beginning of the end of the American passenger pigeon. Once enormous flocks of these pigeons – often consisting of more than a billion birds – populated vast areas of the United States and Canada. The fact that there were so many of them did not keep them from being exterminated. On the contrary, driven together by the thousands, these gregarious birds allowed themselves to be slaughtered for their meat without flying away. In the year 1879 alone the state of Michigan put a billion salted or iced passenger pigeons on the market, and many times that many must have died only to rot on the ground. Only twenty-two years later, on March 24, 1900, the last passenger pigeon living in the wild was shot; afterward, the species lived on only in zoos, where the last specimen finally died in 1914. This species, once the most common bird in North America, was completely extinguished in four decades. The fate of the American passenger pigeon is dramatic, but not unique. Between 1600 and 2000, humans wiped out several hundred species of animals throughout the world. Often, the goal sought was not extinction but the acquisition of leather, fur, meat, fat, or eggs. Sometimes the animals were killed solely for the pleasure of hunting. Passenger pigeons and other "animal resources" perished through a simple misunderstanding: Had humans used them with greater foresight, they might still grace our menus.

In the case of predators, however, the openly stated goal was complete extermination. They were hated as competitors for food and as enemies of people and domestic animals. Humans wanted to eradicate them as quickly as possible. Take the example of Germany: In an order issued in 1697, the margrave of Brandenburg-Bayreuth declared that "both the upper and lower game reserves ... are being ruined by predators, so that hardly any older or younger game is being bagged." In order to improve the situation, it was necessary, the margrave said, "to establish as quickly as possible arrangements that will allow us gradually to eradicate harmful predators, namely: wolves, foxes, martens, polecats, otters, which destroy many fish in rivers and ponds, wildcats, weasels and the like, as well as golden eagles, eagle owls, large, middlesized, and small vultures (various kinds of birds of prey), harriers, small owls, butcherbirds or shrikes, ravens, crows, jackdaws, magpies and jays, and also sparrows that eat grain." The noble ruler wanted in his woods and rivers only those animals that could provide him with a good roast joint of meat – chiefly roe deer, red deer, wild boar, hares – and a handful of game birds. Anyone participating in the hunt for predators was paid a bounty for each animal killed – a proven method that had been used since the sixteenth century to encourage hunters and fishermen to hunt down the unwanted creatures as thoroughly as possible. Gradually, bears, wolves, and lynxes disappeared from western Europe. Otters, beavers, martens, and wildcats were skinned for their pelts, and badgers provided fat for medicines, boot polish, and soap.

Today, European predators are no longer being systematically eradicated. They have survived as species in the wild or in reserves, even if in small numbers. Other species were completely wiped out: In 1627, about sixty kilometers south of Warsaw, a poacher killed the last aurochs. Steller's sea cow – prized for its meat and fat – disappeared twenty-seven years after it was discovered in 1741. The quagga, bontebok, and Burchell's zebra were wiped out by the white settlers of South Africa in order to make room for domestic cattle. We no longer have even a stuffed specimen of the dodo, a bird slightly larger than a turkey that was found on the island of Mauritius. It was first seen by Portuguese sailors in the early sixteenth century, and was extinct by 1681 (giving rise to the expression, "dead as a dodo").

A species is doomed long before its last representative disappears, never to be seen again. Indeed, a few years before the extinction of the passenger pigeon there were still several thousand of these birds, and they were no longer being hunted. Apparently, however, that was too few to ensure the survival of this species, which had specialized in living in huge flocks. How many individuals a species needs in order to survive, what the threshold is that decides whether a life-form will flourish or die out, depends on countless properties – what the representatives of the species eat, and how much and how often; whether they live in groups or alone, and whether they need an individual territory or not; when they reproduce, and how often and with how many partners. Thus a species' survival is not determined solely by the extent of human persecution but also by its biological characteristics.

The gray whales of the Pacific coast of North America have proven to be exceptionally robust. With an estimated population of 22,000 individuals, they are now the most numerous of the great whales. And this despite the fact that since the middle of the nineteenth century this species has been hunted and killed, mainly in

Life and Death
Early hunting was often a very dangerous activity. Nonetheless, Stone Age hunters presumably wiped out whole species of animals.

the lagoons along the Mexican coast, where the whales overwinter to calve. Because the massacre was so thorough, and involved, for the most part, pregnant or nursing females (the calves had no economic value; they were simply abandoned and must have starved to death), by 1895 the species had nearly disappeared along the Pacific coast of North America. But in the second half of the twentieth century it recovered, and was finally struck from the list of endangered species in the United States.

On the other hand, the future of the largest species of whales – and thus of the largest species of animals that has ever lived on Earth, the blue whale – is uncertain. Fewer than 5,000 of these giants – that is, less than 1 percent of their original number – are now thought to live in all the world's oceans. Although this species has been completely protected since 1967, its population continues to decline – perhaps because not enough of them are capable of reproducing. Other species of whales have also been so decimated that they may never be able to recover; the populations of right whales in northern seas have shrunk to a few hundred individuals.

A right whale was killed for the last time off the coast of Long Island in the summer of 1918. One of the whalers later recalled that the catch was scarcely worth the trouble: "We boiled about thirty barrels of whale oil out of it, but we never sold them. It no longer had any market value." Petroleum had replaced the once so avidly sought whale oil. Starting in the sixteenth century – when there was already a shortage of oil in Europe – the oil-rich fat of whales, seals, and seabirds was much in demand, and the need for it steadily increased. The North Cape whale was particularly productive: From a single whale's layer of blubber, which was often as much as fifty centimeters thick, up to 16,000 liters of first-class whale oil could be rendered. In the first half of the sixteenth century, Basque whalers were already taking at least 2,500 North Cape whales every year in the waters off the Atlantic coast of North America. The flesh of these colossi was not used – though it could probably have fed all of Europe. Whalebone, the filtering apparatus of the baleen whale, was more economically valuable, and was used in making corsets, umbrellas, fishing rods, whipstocks, wagon tongues, and even nets,

Cruel Bloodbath
In Roman circus arenas, spectators delighted in bloody battles between humans and animals. The number of animals involved was prodigious.

brushes, and upholstery. However, the most profitable part of the North Cape whale was its oil, which filled lamps and cookstoves and was used in making waxes, soaps, lacquers, and pigments, and in the processing of leather and jute. Whale oil greased the engine of the industrial revolution – until 1859, when the first petroleum spurted from the ground in Pennsylvania.

Yet between the discovery of this new source of energy and the end of industrial whaling in 1988 there was still a long way to go. At first, whale oil was used as fuel for still more rapid and larger factory boats and hunting boats, which took only an hour to render a whale weighing one hundred tons into oil and fertilizer. As soon as one of its species became scarce, the whale-hunting fleets turned to the next largest species. Like all dying industries, the whale hunters worked hard to find new marketing possibilities. And so, in 1956, 10,000 pilot whales were taken in a single season off the coast of Newfoundland – with the sole end of supplying food for mink farms.

Fish in the oceans have not fared any better. At the beginning of the sixteenth century, fishermen were already taking thousands of tons of cod from the waters off Newfoundland, using nothing more than handlines and baited hooks. The exploitation constantly increased, reaching its height in 1968 with a catch of 2 million tons. Shortly thereafter, the whole North Atlantic cod fishery practically went dry; there were no more fish to catch. All the prevalent species off Newfoundland were just as catastrophically overfished, including shellfish, plaice, halibut, yellow-tailed flounder, and witches. Overexploitation has deprived many species of seabirds, seals, and whales of their basic food sources.

Therefore, when one species of animal dies out, it can easily take other life-forms – including humans – with it. This law applies to the brutal way in which American settlers wiped out the Indians in the American West. These Indians' existence depended upon the huge herds of bison, which in 1700 numbered about 60 million individual animals. A hundred years later only half of these remained. "As usual, we took only the tongue, the marrowbones, and the loins, and left the rest lying there," wrote a Mr. Streberg in 1858, regarding his buffalo hunt. Another eyewitness reports meeting a

Earlier Exploitation
Older cultures' techniques of hunting, which now seem simple, were sufficient to drastically reduce the populations of large animals.[1]

Skeletons found at storage sites used by Stone Age hunters		
Predmost, Czech Republic	1,000	Mammoths
Iskaia, Russia	2,000	Aurochs
Istallosköer, Hungary	2,000	Cave bears
Gourdan, France	3,000	Reindeer
Salutré, France	10,000	Wild horses

hunter who "had killed 212 bison in less than three-quarters of an hour." In addition to the tongue, only the valuable hides were taken; everything else remained on the prairie. In this fashion, the Europeans needed only a hundred years to decimate the buffalo population – and along with it the Indian tribes. To honor this murder, the Confederate General Philip Henry Sheridan asked the United States Congress to have a medal struck; it was to bear a dead buffalo on one side – and a dead Indian on the other.

"The earth does not belong to humans, humans belong to the Earth. Whatever we do to the earth affects us in return." Wise words, which Chief Seattle of the Squamish Indian tribe is supposed to have uttered in a message to the United States government in 1855. They warn the "white man" to think about the consequences of his reckless treatment of the environment. As a paradigm of the "red man's" intimate relationship with nature, the speech made its way around the world in various forms during the twentieth century – printed on posters, in environmental magazines, and in children's books. Many people have read these words, but no one knows them better than Ted Perry from Austin, Texas. For it was he who wrote them – for a conservation film made in the 1960s. Modern eco-romantics consider the Indians noble savages and wise conservationists.

The reality was different. The first inhabitants of the Americas often recklessly exploited wild animals and forests; even wiping out species was not an invention of the industrial age. Long before the arrival of Christopher Columbus, the Indians' ancestors had eradicated whole species, using fire, traps, and hunting weapons – at least so claims a group of American scientists. For as soon as these Stone Age peoples came over the Bering landbridge from Asia to North America, 12,000 to 15,000 years ago, the death of thirty-one bird and large mammal species began. These species included mammoths, saber-toothed tigers, and giant armadillos. In North America, this wave of death in the Pleistocene epoch destroyed a full 70 percent of the genuses of all large mammals, and in South America as much as 80 percent of them. Experts disagree as to whether this was a result of Indian hunting or of the warming cli-

Use of animals in the Roman rulers' arenas	
Consul Pompey	used 600 lions in war games and unleashed a rhinoceros on elephants as a special attraction.
Julius Caesar	had 20 elephants fight against men.
Emperor Augustus	had 3,500 animals, including 200 lions, killed in the course of 26 hunting festivals .
Emperor Trajan	had 11,000 wild animals killed during a 123-day-long festival.
Emperor Mark Antony	had 1,000 brown bears slaughtered in a single day.

Lethal Games
In the Roman arenas not only thousands of human beings but also many animals were killed. [2]

mate after the Ice Age. But the original inhabitants of America probably at least contributed to the demise of the large animals.

In any event, in the period following their migration to the American continent, the Indians clearly did not behave like the "noble savages" they are often taken to be. Their rapidly increasing population – by the time Europeans began to settle the continent, they numbered in the millions – had to be fed. Wild animals were a sought-after food source. To get meat, the Indians drove whole herds of hoofed animals over cliffs. Such reckless hunting methods resulted in the extermination in America of archaic horses and camels, and drove mule deer, reindeer, elk, and moose to the edge of extinction. On closer inspection, the image of the prudent Indian whose harmony with nature leads him to use the plant and animal world in a sustainable way turns out, in many cases, to be a myth. These hunting societies' religious rites show how little humans have understood their harmful influence on the nature around them: If game was becoming scarce, they attributed this to the anger of the gods, rather than seeing it as a result of their overhunting.

The fact that other indigenous peoples also decimated, and sometimes even wiped out, their game animals by intensive hunting is shown by the example of moas. Over millions of years, more than a dozen different species of this peculiar flightless bird – they looked like kiwis and emus, but were not related to them – had developed in New Zealand, including the giant moa, which at 3.6 meters in height was probably the largest bird in the world. In the nineteenth century, British colonists plowing the earth turned up the bones of this imposing bird, which was then already extinct. Why had they all disappeared? Until a short time ago it was believed that there was a natural cause for their extinction, perhaps a dramatic change in the climate. Recent discoveries, however, suggest that the Maoris, who arrived in New Zealand about 1,000 years ago, were chiefly responsible for the disappearance of the moas. Over a hundred archaeological excavation sites prove that these first settlers killed moas on a massive scale and cooked them in clay ovens. They ate the meat, made things from the skin, cut fishhooks and jewelry from the bones, and used the blown eggs as vessels. In the known

The Privilege of Hunting
In the Middle Ages, hunting wild game was the prerogative of kings and nobles. Wolves and other predators were seen as competitors for prey and were mercilessly exterminated.

Maori hunting areas alone the remains of 100,000 moas were found. Within 500 years, the Maoris had wiped out the entire genus.

Archeologists found traces of similar massacres on many other Polynesian islands and elsewhere. In Madagascar, the demise of the aardvark, the pygmy hippopotamus, and fourteen different species of lemurs – some as large as gorillas – probably has to be attributed to human hunters as well. In Australia, nine out of ten genera of large marsupials – including a giant kangaroo, a wombat as large as a rhinoceros, and a saber-toothed marsupial – died out after the aborigines arrived from Asia some 60,000 years ago. Many large animals throughout the world are still being hunted so intensively that their survival is in danger. In Africa, traditional hunters and clandestine poachers are devastating many animal populations (including anthropoid apes) in order to satisfy the demand for bushmeat. However, the greatest overexploitation in the present-day world of animals is connected with the production of pharmaceuticals. Tablets, powders, tinctures, and salves made from wild animals according to the recipes of traditional Chinese medicine (TCM) are supposed to heal human illnesses by means of their magical powers. Their true therapeutic value, however, is rarely proven.

Seal penises are used to increase sexual potency, earthworms are supposed to cure nerve spasms, and bears' gall is administered to fight dysentery and jaundice. Cobra blood is said to be good for people with myopia, and if a newborn child's cranial sutures don't grow together quickly enough, try grinding up the undershell of a land tortoise and feeding it to the child.

Tigers are particularly prized for their medicinal potential. In India, leprosy is treated with tiger fat; in Laos, tiger claws are made into tranquilizers, the skin of tigers' ears is used to heal dog bites, and their teeth are supposed to soothe fevers and sore genitals; in Vietnam, tiger bones are a component of a commonly used salve for rheumatism. Tiger testicles are supposed to do wonders for tuberculosis of the lymph glands, and tiger stomachs cure upset human stomachs. Cut up and processed for TCM drugs, a single tiger brings about $325,000. Small wonder, then, that the big cats continue to be taken by poachers despite international

Hunting with hounds
In the eighteenth century, aristocratic hunting was a highly ritualized cultural form. Many large animals were exterminated in Europe at this time.

protection measures: Of eight subspecies, three have already become extinct, and three more are near extinction.

The East Asian drug market spurs poachers on in all parts of the world and has become a lethal threat for many rare animals. Today, the survival of more than 26,000 species is directly endangered by illegal trade. "Chinese pharmacies are sucking up the world's wildlife like vacuum cleaners," says Judy Mills, director of the East Asian office of TRAFFIC, an organization of the World Wide Fund for Nature. In addition to big cats, the saiga and the rhinoceros are particularly threatened. The horns of both these animals are ground up and given to reduce fevers and bring the ill out of delirium and comas. Less conspicuous species are also endangered. For instance, over the last decade the sales of sea horses – used as a cure for headaches and asthma – have increased tenfold in China alone: An estimated 20 million of these peculiarly shaped little fish are taken yearly from the world's oceans. In the waters off Java and Bali its population has sunk by half since 1990.

"If 1.2 billion Chinese suddenly started eating grass, grass would be endangered," says Judy Mills, commenting on the enormous use of animals – and also plants. In fact, a good 80 percent of TCM drugs contain plant materials. Whether they actually relieve suffering or merely heal superstitious minds remains to be seen. In any case, this question is irrelevant to the fate of overexploited species. They are the victims of a false sense of being in harmony with nature; the revival of traditional medicine in East Asia springs from many young people's desire to live in a "natural" way. In this respect, Asians are no different from Europeans and Americans. In Western industrial countries "natural" medicine is also booming. In the United States, a medicine made from yew trees was recently approved for the treatment of cancers of the ovaries and breasts. In order to meet the demand, enormous quantities of bark and needles were required, and as a result, most of the trees are being destroyed. The same thing is happening to the African *Prunus africana*, from whose bark a drug used in treating prostate problems is made. Cameroon is the main country exporting the bark, which is sold chiefly to Italy and France.

Hunting Fur-Bearing Animals Hunters in indigenous Siberian cultures did not hunt solely to provide for their own needs. They also went after sables, whose pelts they could sell to Russian traders.

German consumers are also buying more health care products made from plants than ever before. In many cases, the increasing demand for "natural" salves, pills, teas, and creams does nature no good. Ninety percent of the approximately 2,000 medicinal herbs traded worldwide are not cultivated, but collected in an uncontrolled manner in their natural habitats. Arnica, yellow-flowered gentian, adonis, thyme, oregano – many of the most familiar species – are being pulled, dug, or hacked out of mountain meadows and forests. Thyme is another example – in southeastern Spain, about 75 million thyme plants are pulled up, roots and all, every year. Experts see this not only as possibly endangering the plant's future population, but also as encouraging erosion in the already drought-stricken areas where the plants grow. The modern biobusiness requires mass production; several hundred thousand tons of medicinal herbs with a value of about $12 billion are sold every year worldwide. Germany provides the biggest market for medicinal herbs in Europe, and it is one of the most important hubs for global trade in such plants. More than 45,000 tons, valued at more than $107 million, are imported every year from about 120 countries. In the European Union, sales of these products have doubled over the past ten years. As a result, the survival of at least 150 medicinal plant species is endangered in Europe.

From the Stone Age to the present, hunting has been one of the most important causes of the extinction of species. Yet many species of animals were wiped out completely unintentionally. They perished because human beings were not aware of or ignored the ecological consequences of their actions.

The sailors on the ship *Wellington* had nothing evil in mind when they visited the island of Maui in 1826. Their voyage had taken them from Mexico to the Hawaiian islands, and they wanted to replenish their supply of fresh water. Before they filled their vessels in a brook, they dumped the old water into it. That sealed the fate of the indigenous birds, although this was not known until long afterward. Around 1900 many of the birds living on the Hawaiian islands were found to be dying out – without any obvious cause. Among these were the commonest variety of Hawaiian honey-

Beavers on the Company's Coat of Arms
In 1670, the Hudson's Bay Company began operations in North America. In only 135 years, trappers delivered 16.5 million beaver pelts to buyers alone.

109

creepers. It was a long time before the riddle of their disappearance was solved: They had been carried off by a special form of malaria. Central American mosquitoes, stowaways in the *Wellington's* drinking water, were the vehicles by which this previously unknown tropical disease made its way into the South Seas. The birds were not the only victims of this epidemic; the days of five species of the hibiscadelphus plant, which grows only on the island of Hawaii, are also numbered. Its narrow, tubular flowers are so bent that they could be pollinated only by the Hawaiian honeycreeper, with its specially-adapted curved beak. Inhabitants of islands – whether plants or animals – are in particular danger of dying out, because many of them exist only on "their" island and nowhere else. If they disappear there, the whole species disappears with them. On Madagascar whole species of tenrecs, almost all amphibians and lemurs, more than 90 percent of the reptiles, eight out of ten plant species, and half the birds are endemic – that is, live only on this island. There are many endemic species on all islands that are far enough from continents to shelter their populations from outside influences over a long period of time. Over millions of years, these populations can adapt themselves to the special conditions of their often unusual habitats.

This specialization puts island inhabitants at a crucial disadvantage; having been cut off from the rest of the world for such a long time, they are unable to defend themselves against dangers that suddenly appear from without. As they have evolved on islands, many insects and birds have given up flying and relied on their legs, including giant cockroaches, giant crickets, owl parrots, and even a flightless nocturnal moth. In their isolation, they apparently had no need to fly away from anything. Each organism takes the path of least resistance: Where there are no enemies, animals have their young in unprotected nests on the ground, and plants do without poisons and thorns.

Because remote islands lack experience with invaders, any contact with foreigners can end in catastrophe. Domestic cats and dogs hunt down unsuspecting flightless birds and clean out their unprotected nests. Mongooses, which were introduced into the Carib-

Sad Statistics Whaling statistics in the twentieth century clearly show the results of excessive hunting. When the blue whale became rare, the whaling fleets turned to other species. 3

Number of Whales Killed in Antarctica			
	Blue whales	Humpback whales	Sperm whales
1920–1930	72,472	2,866	465
1930–1940	161,155	11,059	8,129
1940–1950	40,021	5,222	12,214
1950–1960	28,168	13,134	49,091
1960–1970	3,946	13,000	43,318

bean islands to fight rats, killed anything they could catch. The rats themselves, which are found in every place where ships land, are also devastating for the local fauna. Thirty to fifty-five species of seabirds in the Pacific tropics raise their young only on islands that are free of rats – and there aren't many of those left.

So humans have nearly eradicated countless species of plants and animals by unleashing on them – partly intentionally, partly unintentionally – competitors and enemies. Like hunting, this indirect way of destroying nature has a long history. More than 1,000 years ago, the earliest humans to reach Hawaii brought with them from their Polynesian homeland sweet potatoes, dogs, chickens, small pigs, and a tropical species of rat. Captain Cook, and other European seafarers of the eighteenth century, stocked the South Sea islands with cattle, pigs, sheep, and goats. Horses, donkeys, chickens, and cats soon followed them, along with food plants, from pineapples to sugarcane. In time, some of the domestic animals went wild and ate up the native vegetation. Many birds caught fowl pox from the chickens.

Today, foreign intruders are increasingly threatening animal and plant worlds, not only on islands but also on continents and in the water. In the United States, they are the chief threat to 40 percent of the endangered species of animals and almost 60 percent of the endangered species of plants. Mollusks get into the ballast water of trading ships traveling from the Black Sea to Europe and to the Great Lakes in North America; microscopically small poisonous algae travel from Japan to Australia; in Hamburg's harbor, longshoremen discovered piranhas, and Chinese crabs of the species *Eriocheir sinensis* have been found in the North Sea.

From time immemorial, species have always sought to extend their habitats. Plant seeds fly over mountains and seas, land animals travel thousands of miles and are used as buses by insects and other small animals. This natural process has always resulted in newcomers breaking the food chain and disturbing whole ecosystems. But human mobility has significantly accelerated this process. As stowaways on ships, airplanes, and cars, species move from one continent to another at high speed. The consequences of this biological globalization are not yet foreseeable.

Whaling
Even when whalers still had to throw their harpoons by hand, they killed thousands of whales.

Many foreigners adapt to their new biotopes in an inconspicuous manner. This is the case for the American raccoon in Europe. But other experiences remind us that we must be cautious. In the nineteenth century, 200 to 300 species of cichlids were wiped out by Nile perch that had been introduced as a food fish. Foreign plants as well as animals can upset interconnected ecosystems. Tahiti provides a dramatic example. In 1937, *Miconia calvescens*, a tall species of myrtle that grew in South American rain forests, was introduced as an ornamental plant because of its attractive leaves. Today, it has taken over two-thirds of the island – from the dry areas along the coast to the misty forests in the mountains. This outsider is threatening 40 of the 107 endemic plants, deprives ground-covering plants of light, and thus makes the earth more susceptible to erosion and avalanches.

"There is now, in comparison with what was earlier, only the skeleton of a body, which is consumed by illness: All around the fatter and softer earth has washed away, and only the meager bones of the land have remained." This complaint could be made in our own time by a farmer in Madagascar or a shepherd in the Sahel. However, it was made by Plato, describing the decline of his Greek homeland about 400 B.C.E. Humans, and not, for instance, climate change, were mainly to blame for the erosion that turned the fertile Peloponnesus into the dry, barren land it is today. For hundreds of thousands of years, humans have been setting fires to clear woods and grasslands; for the past 10,000 years they have been driving out wild plants and animals in order to make room for their settlements, farms, and pastures. Nature is what we make of it: The Scottish highland is now covered with heather because our ancestors cut down the oak and pine forests that once stood there. The Australian aborigines not only destroyed the large animals, but also wiped out the vegetation of their continent by regularly setting fires. Easter Island was covered with a subtropical forest until humans began plundering this flowering paradise some 1,500 years ago. Within nine centuries they had reduced it to an uninhabited steppe, in which they themselves ultimately starved to death.

Loss of Forests Since 1940, the forests of Costa Rica have been cut down at an unusually rapid rate. Many of the remaining forests are now protected.4

However, the sins of the past pale in comparison with the speed and extent of our current alteration and destruction of natural habitats. This meteoric process is now the chief threat to biodiversity on our planet. It is the leading cause of the extinction of species – far ahead of invaders, hunting, and environmental pollution. The growing world population needs more and more space, food, and energy. In developing countries, where more than 80 percent of all people now live, about 400,000 square kilometers of forest land were transformed into pasture and farmland from 1973 to 1988 alone.

This affects not only rain forests but all kinds of habitats: mangrove forests, steppes, savannas, and marshlands. Of the originally more than one million square kilometers of prairie land in central North America, only fragments remain – and one can say the same about the South American pampas and the steppes of southern Russia and Ukraine. In the Indonesian–Malaysian area and in Africa, less than half of the mangrove forests that once lined the coasts of tropical and subtropical seas and served as a nursery for fish and other inhabitants of the sea still remain.

What is not being intentionally burned or destroyed with bulldozers is falling victim to the side effects of modern agriculture. Most of the endangered species in Europe are becoming rarer and rarer because they cannot endure the excessive fertilization of agricultural lands, the monotony of agricultural landscapes, and the use of pesticides.

By 2050, the world population – according to the UN's current predictions – will reach about 8.9 billion people (a medium scenario). It will need more millions of square miles of land for cattle and for wheat, corn, and rice fields, for timber plantations, and for settlements – and this means that still more species will be deprived of their habitats. IUCN plant expert David Given predicts that if the destruction of natural habitats continues at the present rate, 34,000 flowering plants – that is, one out of eight of the species that have been described so far – may disappear forever. In many climatic regions – for example, the south tip of Africa – the future of a third of all the flowering plants is at stake. Such numbers are crude estimates – and they take

Lost forever
Since 1600, 611 animal species and 396 plant species have disappeared, largely as a result of human influence. Most of them died out in the eighteenth and nineteenth centuries. Awareness of this loss first emerged in the twentieth century. [5]

into account only the species that are already known. How many more of the existing but unknown species of animals and plants, fungi, and microbes are threatened, no one knows. We don't even know how many species there are on our planet: 5, 10, or 100 million?

Biologists agree on at least this much: Most species – estimates range from 50 percent to 70 percent of all animal and plant species – live in the tropical rain forests. About 50,000 different species of trees live in these forests alone. In the crowns of a single species of tree in Panama, biologists found no fewer than 1,200 different species of beetles. Biologist Edward O. Wilson counted forty-three species of ants on a single tropical tree. Experts disagree as to how many species are endangered by the destruction of the rain forests. But one thing is certain: The destruction of tropical rain forests and other habitats deprive many species of insects, spiders, fungi, and other small living creatures of their very existence.

Since the extinction of the passenger pigeon, it has become clear that a species can be seriously endangered even when it still appears to have a sufficient number of representatives. A species' genetic diversity is crucial to its survival. The more varied the properties of the individual members, populations, or breeds of a species are, the more vital is the species as a whole, and the more capable it is of adapting to altered living conditions. Because humans are destroying more and more natural habitats, many species are reduced to small residual areas that lie amid farms, pastures, and settlements, like islands in a sea. Imprisoned in its isolation, each small group of creatures has access to only a small fraction of its genetic heritage. In the IUCN's view, this impoverishment of the gene pool represents a serious threat to many species.

On the other hand, the practice of international conservation offers examples of the successful resurrection of species that had already been reduced to a few individuals. The northern elephant seal, the Père David's deer, the wisent or European bison, and other rare species have multiplied a hundred- and a thousandfold, even though they all descend from a handful of ancestors that barely survived extinction. And this despite the fact that the genetic diversity of these species is almost nil.

The loss of biodiversity concerns not only wild life-forms, such as gorillas and orchids, but also domestic cattle, rice, and many other animals and plants that are used as food sources. Over the past 10,000 years, farmers all over the world have developed 5,000 breeds from only three species of animals. Yet in many developing countries, breeds of chickens, pigs, and cattle that were adapted to local conditions are now being replaced by standardized, high-production animals. However, in Europe as well, farmers have abandoned many breeds and varieties since the turn of the twentieth century. Of the 2,238 remaining European breeds of animals, half exist only as small remnant populations. Many kinds of the most important cultivated plants have disappeared just as quickly; since the beginning of the twentieth century, the genetic diversity of grain varieties has decreased by 75 percent. The work of whole generations of breeders and farmers is being lost. Experts working for the United Nations' Food and Agriculture Organization (FAO) warned years ago that the food supply for future generations was being seriously threatened by the loss of genetic diversity.

We can at least rightly take pride in one campaign of destruction: the worldwide, final extermination of the smallpox virus. On October 30, 1997, the American Centre for Disease Control and Prevention (CDC) celebrated the twentieth anniversary of the victory over an enemy that killed 2 million people a year and permanently damaged even more. But ironically, in this of all cases, humans have made an exception: The planned destruction of the last smallpox virus, which is stored in frozen form in American and Russian laboratories, has been delayed.

See also plates:

A Changing World
Page 24
Hot Spots of Biodiversity
Page 32
Globalized Nature
Page 34

Sources for text in margins:

1 E. Hobusch: Das große Halali, Berlin 1986
2 Ibid.
3 R. Ellis: Mensch und Wal, Munich 1993
4 WCMC
5 A. P. Dobson: Biologische Vielfalt und Naturschutz, Heidelberg 1997

Threatened and Threatening: Our Love-Hate Relationship with Nature and Its Conservation

Guest author: Jeffrey A. McNeely, Chief Scientist, IUCN (The World Conservation Union). Humans have a love-hate relationship with nature. On the one hand, they venerate it. On the other, they fear its dangerous aspects: man-eating tigers, malaria-carrying mosquitoes, marauding elephant herds.

Colombo, the capital of Sri Lanka, the green island off the south coast of India: People are marching in large crowds through the streets, demonstrating for a great cause. Today the cause is not peace or social justice but, rather, protection for elephants. In Sri Lanka, elephants are an age-old cultural symbol of immense significance. Yet most of the demonstrators are city dwellers. For them, the fact that in the preceding day, two girls were reported killed by an elephant while they were walking in the forest is of no more importance than the report that two days earlier, workers – young, adult, and elderly – died when an elephant trampled their hut on a sugarcane plantation.

We are always ambivalent about nature. We are completely dependent on what nature provides and does for us, and we need nature spiritually and mentally, as well, but we are also threatened by animals as carriers of disease or as hungry predators. Although many large mammals – elephants, rhinoceroses, lions, and wolves – can kill humans, many people are working to save these very species.

In the course of the twentieth century the world population has grown from about 1.5 billion to about 6 billion. The per capita consumption of natural resources has also increased, and this means that we are leaving a correspondingly smaller amount of these resources for the other life-forms with which we share the planet. Whereas sources of disease have been able to multiply greatly as a result of the extension of the human habitat, and cockroaches, pigeons, sparrows, and rats have thrived in the wake of civilization, the populations of other species have significantly decreased. Some have already died out, at a very disturbing rate – at least four times as fast as the natural rate at which species were lost in earlier ages. If we examine historical descriptions more closely, it seems that humanity can't help causing the extinction of species, or even whole genera.

However, the effect of human action on plants and animals differs, depending on the region of the world. Whereas in Africa or Asia, environmental changes have been accompanied by a relatively low quota of extinctions, areas only recently settled by humans have suffered much more. As we have already noted, after humans first crossed the Bering landbridge from Asia 12,000 to 15,000 years ago, more than 70 percent of the large mammals died out in North America, and in South America, 80 percent. After humans reached the Australian continent, about 60,000 years ago, 86 percent of the large mammals died out. It is true that Hawaii and Great Britain have about the same number of bird species that breed on land (about 135). Yet on the Hawaiian islands, which were settled a little over 2,000 years ago, more than 100 species have already died out, whereas Great Britain, which was settled many thousands of years earlier, has lost only three species over the same period.

Australian biologist Tim Flannery has suggested that the history of humans in Africa, Asia, and Europe "reads like the story of a cat and a bird. ... In contrast, the history of humans in 'newly' settled lands is more like the story of a cat and a bird that have grown up together. Familiar with each other from the beginning, they live together in peace. In contrast, the history of humans in new lands is more like the story of a cat and a bird in which the cat has grown up among other cats and the bird has never seen a cat before."

Hunters often have a significant impact on wild animal populations, as is shown by the numbers of prehistoric extinctions mentioned above. Not long ago a scientist discovered that toward the end of the 1980s, up to 19 million mammals, birds, and reptiles were being killed every year in Brazil's Amazon region. The total number of all killed or mortally wounded animals may be as high as 57 million. In the Peruvian province of Loreto alone, people kill about 370,000 monkeys a year, either for their own consumption or for sale.

In the eighteenth and nineteenth centuries, large predators were exterminated in broad areas of their range, especially when there were direct conflicts with humans. In the Alps, brown bears, wolves, and lynxes were increasingly driven out by logging, hunting, and the declining populations of the animals on which they preyed, so that at the beginning of the twentieth century they were practically extinct.

Among birds the losses were particularly severe. In the tropical Pacific islands, humans had already wiped out numerous species over a period of several thousand years. Bones excavated show that most land and seabirds had already disappeared in prehistoric times:
- Over 2,000 species of birds are supposed to have died out in the South Pacific as a result of human influence. Most of them were flightless rails.
- The number of species of seabirds nesting on Ua Huka (the Marquesas Islands) has sunk from more than twenty-two to four; on Huahine (Society Islands), where there were more than fifteen species, there are also now only four.
- Between the time the first humans landed on Hawaii 2,000 years ago and Captain Cook's arrival in the islands a little over 200 years ago, sixty indigenous land-bird species (of which we know only from their bones) died out, and twenty to twenty-five additional species have died out since the Europeans' arrival.
- Bones found on New Zealand's North Island show that twenty land-bird species were lost in prehistoric times, and seven more that still existed when Europeans arrived there have since become extinct. Of fifty land-bird species that once lived on New Zealand's South Island, more than half have disappeared.

Altogether, such figures amount to a 20 percent reduction in the number of bird species worldwide.

Freshwater animals are even more seriously endangered than animals living on land. In North America, 36 percent of the river crabs, 55 percent of the mussels, and 20 percent of all fish have died out or are at least endangered (compared to only 7 percent of birds and mammals). In East Africa's Lake Victoria, some 200 fish species which exist nowhere else were lost forever between 1980 and 1990. Since 1990, an additional thirteen species have disappeared along the banks of the lake; according to a long-term study, the causes of this extinction are fishing and increased nutrient content, exotic species of fish introduced by humans, and a fast-growing water hyacinth imported from South America.

As a result of world trade, pathogens are transported from one continent to another with alarming frequency. Because of this mobility and the speed with which many germs develop, it is very likely that animal populations will be increasingly threatened by unknown diseases. Epidemics are probably the cause of the clear decline of at least fourteen frog species that live in the rivers of the rain forest in the mountainous part of eastern Australia. Frogs in Africa, North America, and Europe are also affected. An irony of fate – at least one of the diseases was transmitted by the very scientists who were studying the extinction of frogs.

For thousands of years, humans have had a devastating effect on wild animal populations. A combination of various factors – new technologies, the adaptation of land to human needs, the rapid rise in human population, and increasing consumption of natural resources – gave people the power to wipe out plants and animals on a massive scale.

In their struggle for survival, *Homo sapiens* has had to fight countless battles. For instance, in Europe and Asia man-eating wolves were a very real problem:
– In France, there was a wolf called "the Beast of Gevaudan" that devoured sixty people before it was finally killed.
– In 1712 almost a hundred people were killed by wolves in the forest of Orléans in France. According to experts, France suffered

more from wolves than any other country (the last known case occurred in 1914). Fairy tales like "Little Red Riding Hood" were written, in which children are the chief victims of wolves.

– As late as January 1968 wolves killed eighteen people near the Iranian city of Hamadan.

In Russia as well, this four-legged predator was always a special danger, taking a bloody toll of 200 humans a year. In December 1927, wolves besieged the Siberian village of Pilowo. First, they killed all the guard dogs, and then began dragging women and children out of the houses that lay on the edge of the hamlet. It was reported that wolves ended up breaking down doors and attacking villagers in their houses. Finally, army troops were sent in to prevent the wolves from wiping out the whole village.

Big cats usually do not regard humans as prey, but rather as competing predators; nonetheless, they continue to attack humans. In 1906, while a bridge over the Tsavo river in Kenya was being built, over a period of nine months, two lions killed twenty-eight Indian and a dozen African workers. More recently, a lion in the Luangwa valley in Zambia killed fourteen people in a single month, sometimes tearing down the doors of huts in order to get at its victims. In Rudyapryag in India, a man-eating leopard killed 125 people between 1918 and 1926, when it was finally killed by the famous big-game hunter, Jim Corbett. On another occasion, Corbett shot a leopard that was supposed to have killed 400 people.

Whereas the number of wildcats decreases in Asia and Africa, the number of pumas in North America seems to be growing, and increasingly this leads to dicey situations. In California, puma hunting has been banned since 1972, and as a result, the population of these cats has since doubled and is now about 5,000. Until 1990, pumas killed only one person per decade, but in 1992 alone they killed two California hikers and two sportsmen in Colorado.

Tigers are the most dangerous wildcats. Over the past four centuries, they are estimated to have killed a million people, or 2,500 per year:

– In a village in the northern part of modern-day Myanmar, a tiger killed twenty-four people in only four days.

– In the Sundarbans mangrove forests of India and Bangladesh, more than 600 people were killed by tigers between 1975 and 1985.
– Between 1978 and 1987, 170 people died in India's Dudhwa National Park. Tigers live there on a reserve that borders directly on the sugarcane plantations. The tigers like to use the fields for their excursions.

Tigers that kill people usually have been wounded in some way. That is why they attack humans rather than their usual prey; humans are numerous, easily available, and easy to kill.

Interestingly, many people who live in close proximity to tigers often feel a kind of reverence for them. This is probably because tigers prey mainly on animals that damage crops, such as deer, wild boar, and monkeys. For many villagers, these powerful beasts are wreathed in myths. They are seen as divine avengers that punish only those who have broken tribal taboos. In many areas of Sumatra, it is believed that Allah has entrusted the tiger with seeing to it that sinners receive condign punishment. When in 1951 one hundred people were killed on the southwest coast of Sumatra – not far from the city of Bengkulu – this was interpreted as Allah's vengeance, and the tiger as his executioner.

The ticklish cohabitation of humans and tigers is changing because more and more people are settling in the forests and savannas, where once the big striped cats were absolute rulers. Any tiger that steals cattle or kills villagers must now pay for it. Yet in areas where tigers have become a serious threat, hunters take care to kill only the man-eaters, not their harmless fellows (at least where tradition still holds sway to some extent).

And then there are the "small predators." Rats are one of the few animals whose populations are rapidly growing – not despite, but rather because of, the fact that humans are constantly extending their range. It is estimated that more than 40 million rats live in Yangon, the capital of Myanmar. In India, there are at least ten rats for every human inhabitant. The 4 billion rats estimated to live in China eat more than 15 million tons of grain every year.

Humans increase rats' quality of life, even though we may not mean to do so. But the rodent still proves to be an ungrateful guest.

Some thirty diseases which are transmitted by rats have cost more human lives than all wars and revolutions taken together, and the typhus they carry has put a larger mark on history than any political leader has. Among the many serious diseases carried by rats are trichinosis, rabies, leptospirosis, murine typhus, scrub typhus, tularemia, and salmonella. Many of these diseases are most dangerous in urban areas, where humans and rats live in a relatively confined space – an explosive combination.

Now, as earlier in history, the most feared, and probably also the most dangerous, threat is bubonic plague, the Black Death. Carried by a flea that lives in the fur of rats that live among humans, in the middle of the fourteenth century it swept over all of Europe and then throughout Asia. By 1351, it had already killed an estimated 25 million people. Experts think that more than a third of the world's population died during this period. In 1665 the plague raged again in London, killing almost 75,000 of the city's 460,000 inhabitants.

Humans have always waged war on any insect that destroys or ruins food plants in the field or in granaries, attacks our domestic animals, or conveys deadly diseases. However, the weapons used in this war have changed; from simple but effective crushing and swatting, we have moved on to spraying synthetic insecticides from helicopters and manipulating genes to make grains immune to insect attacks.

Each time, technological innovations have promised to put a decisive end to the battle against our tiny adversaries, but each time, the latter have had the last word. While we have made great progress in conquering our habitat, replacing whole interconnected plant associations with agrarian ecosystems, and covering the earth with asphalt in order to facilitate transportation, in the battle against insects we have gained hardly any ground. They still destroy a large part of our crops, invade our houses and offices with impunity, and spread diseases.

It is highly probable that the animal most dangerous to us is the Anopheles mosquito, which carries malaria plasmodium, a pathogen that has already killed millions of people. In the tropics, hundreds of thousands of new victims are infected every year.

Researchers have recently discovered that the viral disease AIDS also originated in the realm of animals; it was transmitted from chimpanzees to humans. In at least three regions of West Africa where chimpanzees were eaten, the virus moved to a different host species. This kind of transfer to us through hunting presumably happens fairly often, although it seldom results in epidemics. In West Africa, chimpanzees are hunted for their meat, which is sold in the cities or to supply logging camps with food.

The smallest predators, from viruses to rats, are an important part of our ecosystem. We will never be rid of them, no matter how hard we try. Instead, we should better adapt to them and develop a broad spectrum of countermeasures to limit their negative effects. One possibility consists in maintaining a certain balance in the ecosystem; species that see us as prey should be kept in check by other predators.

For example, where rats damage crops, an intact owl population will help. Dangerous mosquito populations can be kept down by improved water management that promotes the proliferation of dragonflies (which eat mosquitoes) and frogs (which eat mosquito larvae). In short, the maintenance of a many-faceted and healthy ecosystem is surely the best way to maintain a balance between humans and the rest of nature.

From mosquitoes to elephants: Even pachyderms, at the other end of the scale in size, can become pests, particularly when humans dispute the forests with them. The French colonial rulers in Vietnam often gained the impression that elephant herds sensed the danger of approaching civilization. The animals destroyed all the telegraph poles, bridges, and isolated forest outposts they could find. They tore down miles of telegraph wires and tangled them so much that it took repair teams days to untangle the cables and restore connections. "Hardly was everything repaired than it was destroyed again," reported William Baze, who was entrusted with the difficult task of pushing ahead with the development project in the Vietnamese forests. "In the course of a month, this constant cycle of destruction, repair, destruction, repair became unbearable. Railings were twisted or ripped out, milestones along the roads were overturned or moved, forest shelters lost their roofs and doors, barrels of cement and tar were rolled over the roads and hidden everywhere – until finally no one found it funny anymore and they all got sick of it." Baze settled down to wait, and shot some of the biggest elephant bulls that had been terrorizing his project. Apparently this sufficed to frighten away the rest of the elephants long enough to successfully complete the work.

In 1983 several elephant families were driven out of the Hosur forest area in southern India; they retreated in the face of expanded farming and moved north to the state of Andhra Pradesh, where wild elephants had not been seen for more than a hundred years. There they destroyed fields and killed several people who did not know how dangerous it can be to come too close to an elephant.

In November 1993 a herd of about fifty elephants moved toward Calcutta, leaving a corridor of devastated rice fields behind them. When villages and the local authorities made a concerted stand against the elephants, they eventually turned around. In all these cases it was ultimately people who had destroyed habitats and thus provoked the dangerous encounters. For under normal conditions, forests and steppes are far safer than cities, where dangers of all kinds – from automobiles to hostile humans – present a far greater threat than any animal one might meet in the wild.

The World Bank has expressed concern about the damage done by wild elephants to its projects for transforming forest areas into plantations. A 90-million-dollar palm oil plantation in Malaysia suffered greatly from pachyderms that were not satisfied with eating the palms, but walked along the rows and systematically trampled every tree. Two million seedlings were destroyed, and the project's cost turned out to be 94 percent higher than planned.

In Sumatra as well, the conflict between humans and animals came to a head because the people were settling farther and farther into the natural area. Whole villages were razed by attacking herds of elephants because immigrants from overpopulated Java had simply appropriated the elephants' habitat.

What can we do in this kind of conflict situation? I lived for twelve years in areas of Asia inhabited by elephants, seeking an answer to this question. Elephants are highly intelligent animals, and it can be assumed that they also are trying to find solutions. How can we harmonize our interests with those of elephants? Some forms of land use are absolutely compatible with the animals' needs, such as forestry (unlike farming, which is, in any case, not very viable in the hilly areas often covered by forests). Since every country needs productive forests, one way to begin would be to manage them in such a way that they could also serve as a habitat for elephants. This requires long-term planning, and logging should be done carefully. Forests managed in this way could have advantages for both sides; they would provide the elephants with sufficient room and humans with a significant economic resource.

There are a few hopeful signs that even large predators can have a place in our modern world. Alpine forests have extended their range in recent years, hoofed animals living in the wild have further increased in numbers, and the human population is increasingly accepting the gradual return of large predators, as well as of other animals. In the 1960s, lynx were already being sighted in the Alps again. In the early 1990s, wolves moved from central Italy back into the southwestern Alps. The brown bear has migrated from Slovenia back to the Austrian Alps.

The return of these animals to their former environments is important and offers an eloquent perspective on conservation. Recently, experiments in reintroducing wolves into the wild have been carried out, and the results have clearly shown what a positive influence large predators have on the ecosystem. In 1995 and 1996, thirty-three wolves were released in Yellowstone National Park. At the end of 1997 they had increased to ninety-seven in number. In two years they killed half the coyotes living in the park, forced the elk to be more alert, and saw to it that carrion-eating species increased their numbers because they fed on the wolves' prey. Since there are fewer coyotes, small rodents have spread, which is a boon for birds of prey. On the whole, biodiversity has greatly increased.

The opposite case – extinction – also offers indications of how important an ecological role is played by large predators. When pumas, jaguarundi, and ocelots were driven out of the ecosystem on the island of Barro Colorado (Panama), grave consequences ensued. Because small, egg- and chick-eating predators such as coatis, raccoons, and opossums greatly increased in numbers, several species of birds that nested on the ground disappeared.

The destruction of habitats about which conservationists repeatedly complain is often only temporary. For example, forests in the eastern part of the United States are constantly expanding (three-quarters of the forested areas of the country are now in the East) and the associated fauna are also increasing in number. White-tailed deer have become so numerous, for instance, that they collide more and more frequently with automobiles: In 1990, about 43,000 deer were involved in accidents in Pennsylvania alone. They so often eat young shoots in gardens that many homeowners call them "rats with hooves."

About 40,000 black bears are again roaming the eastern United States. Alligators were put on the list of endangered species in 1967, and hunting them was prohibited; within ten years, their numbers swelled to nearly 2 million. In 1972, thirty-seven wild turkeys were released in western Massachusetts, and now once again populate – 10,000 of them – an area in which the species had long ago died out. Consequently, predators, such as coyotes, that never before lived in Massachusetts have settled there.

At the turn of the twentieth century, moose were nearly extinct in the United States, partly as a result of hunting, and partly because the forests in which they lived had been cleared for farmlands. However, a series of events led to a population explosion – tight restrictions on hunting, conversion of farmland back into forests, and the decimation of their chief predators (wolves). In addition, there was the spread of a new form of forest management that favored young trees whose leaves grew at a low level and so provided the moose with sufficient forage. At one time there were moose only in northern Maine; now they are found in central and southern New England and even in eastern Pennsylvania. Today, there are about a million moose – three times as many as in 1945. And where moose go, wolves soon follow.

However, modern legislation promoting the preservation of species is not always supported by the populace. In many farming areas in the Alps, large carnivores continue to be regarded as murderers of wild game and cattle. On the other hand, in the Alps the ecological conditions and farming have fundamentally changed since large predators died out there, so that the potential for conflicts has correspondingly declined.

All the same, cattlemen have lost the traditional sense of sharing the habitat with large predators. Moreover, there are again a great many sheep in the Alps. The return of large predators will not be possible without basic changes in the whole system of sheep raising. At present, the local population is not willing to take such steps, since they reject any outside influence on their lifestyle. They consider large predators as a negative symbol for restrictive conservation, which, in their view, hinders economic development.

Thus, any successes in conservation are accompanied by new demands. Protecting tigers and elephants leads them to spread, and this results in more conflicts with humans. The return of predators to their former habitats irritates those who have become used to their absence. To this extent, the protection of wild animals requires three steps, which complement each other. First, informational campaigns must encourage greater understanding of the importance of animals living in the wild and of contact with them on the part of both the government and the general population. Second, appropriate land-use planning should extend wildlife reserves and buffer zones, restore and maintain areas for animal population expansion, and limit human access to sensitive habitats such as riverine forests. Third, resource needs of people who live near these protected zones must be brought into harmony with conservation; for instance, villages could be accorded economic advantages in return for performing a kind of guardianship of the wild areas and of the wild animals that people both fear and revere.

See also plates:

Human Favorites
Page 18
Humans as Habitat
Page 28
Humans as Victims and Prey
Page 30
Globalized Nature
Page 34
Protected Areas: Four Countries Compared
Page 36

Progress through Catastrophes: How Extinctions Further Evolution

Guest author: Professor Josef H. Reichholf, Scientist at the Zoologisches Staatssammlung München (Munich) and member of the board of directors of the German World Wide Fund for Nature (WWF). There is no equilibrium in nature – except during the short pause between two disasters.

Before humans attacked and altered nature, 0.000009 percent of all living species on Earth died out every year. This is generally regarded as the natural rate at which species are lost. (American paleontologist David M. Raup provided this figure in a report to the United States National Academy of Sciences.) How many extinct species per year or per century this represents in absolute numbers depends on the number of total species. Raup assumed a total of about 2 million species. On this assumption, every five years a species disappears as a result of natural processes. Is that too many or too few? And what does this rate of extinction mean for nature?

It is clear that 99 percent of all the species that evolution has thus far produced have fallen victim to a natural process of extinction. The splendid and impressive biodiversity that still exists therefore represents no more than a tiny remainder of all the life-forms that have existed on Earth. And yet the planet has perhaps never had such a wealth of species as it now has (with the exception of the last quarter of the Tertiary period, about 15 million to 3 million years ago, when the Great Ice Age began). This situation results from the extinction of some species and the emergence of others. The balance between the two processes describes the dynamics of evolution. Millions of fossils that have been collected and studied throughout the world over the past two centuries reveal these dynamics and prove that life never really came to a stopping point. A span of 500 million years can be described on the basis of the fossils discovered (fossils that allow the eras of the earth to be determined). This length of time produced a constant increase in global biodiversity, interrupted by short phases of mass extinction. The largest of these occurred about 250 million years ago, at the end of the Permian period, when, in a short time (by geological standards) 52 percent of all existing families of marine animals died out. Extrapolating to the number of species, paleontologists estimate that at the end of the Permian period, the total loss in marine species was between 77 percent and 96 percent.

A similarly massive extinction occurred 65 million years ago and put an end to the age of dinosaurs. There were four or five other great events of this kind, and between them many smaller ones.

Toucan

Catastrophes – great declines in the diversity of life – mark the stages in the evolution of species. The average life span of a species is a mere statistic. It is estimated to be from 1 million to 10 million years; but what does that mean when a sudden catastrophe – the impact of a giant meteor or an enormous volcanic eruption – puts an end to a species before its time? The survival of the fittest? Or does chance dictate which species survive and become the starting point for new developments? What is the result of all the catastrophes that are so impressively documented by the massive extinction of species?

The interest of these questions is not limited to geological history. They are the basis for the way we deal with nature and its diversity of species. These questions are involved in our ideas of the economy of nature and the meaning of species within it. And they also have to do with the possible consequences when humans cause species to die out. How is the human catastrophe to be assessed with respect to the survival of biodiversity?

Of the four or five largest catastrophes in geological history, the most recent is the best documented. In all probability, it began 65 million years ago when a heavenly body collided with the earth – a body more than ten kilometers in diameter. It struck planet Earth at the edge of the Yucatan peninsula in the Gulf of Mexico. The collision produced a gigantic tidal wave and a clouding of the atmosphere that lasted for years, dimming the sunlight necessary for life and causing sudden a drop in Earth's temperature. It may also have caused volcanic eruptions elsewhere as the shock wave from the impact moved through Earth's core. It is hard to imagine such a change which not only killed off almost all animals weighing over ten kilograms, but also, for a long time, affected the composition of the plant world and the living conditions in the seas. With this event the Mesozoic era ended, and with it the age of the rule of dinosaurs. A new time period began, the Cenozoic.

The catastrophic extinctions nonetheless opened up new possibilities. Surviving life-forms that had hitherto been insignificant, marginal figures on the scene of life, suddenly became central to what was happening and to future developments. The new era belonged to birds and mammals. Both groups had already been in existence for a long time; their beginnings date back to the age of the earliest dinosaurs. But for many millions of years, they remained unimportant and inconspicuous in comparison with the giant reptiles. There is no reason to suppose that they played any role before the great catastrophe. Had they simply died out after they had lived for 50 million years or so, nothing would have suggested their later success. We humans would certainly know nothing about them – we wouldn't exist!

As strange as it seems, we owe our existence to this catastrophe and other catastrophes that occurred during the Great Ice Age. The great turning point 65 million years ago gave all warm-blooded birds and mammals a decisive advantage: The earth was now largely free of the very efficient reptile families. Within a few million years a genuine explosion of new species occurred. Special adaptations evolved that conquered every part of the earth. The ancestor of all mammals looked like a shrew. From this simple form developed elephants, whales, hoofed animals, cats, marsupials, and apes – and from them, humans.

Puffin

How Many Are Lost?
Experts' estimates of rates of extinction vary by as much as a factor of fifty.

Author, Year	Estimated Rate of Extinction	Loss per Decade in %
Myers 1979	A million species between 1975–2000	4
Ehrlich & Ehrlich 1981	50% of species up to c. 2000, 100% by 2010–2025	20 – 30
Raven 1988	25% of species between 1985 and 2015	9
Reid 1992	2–13% of species between 1990 and 2015	1 – 5
Wilson 1993	0.2% to 0.3% of species per year	2 – 6
Mace 1994	Vertebrate species on the endangered list (IUCN)	0.6 – 5

The same can be said about the development of songbirds (Plo-ceids), a family within the order Passeriformes. Today, they represent two-thirds of all bird species on Earth, and have far outstripped all other branches of the bird family in variety. Over millions of years, the plant world has also developed a splendid diversity, on which the immense variety of insects is dependent. Every new development leads to further developments. Diversity builds on diversity; the more it thrives among plants, the more diverse the animals that live on them can become.

That is the good side of catastrophes, which also appear – at least on a small scale – in our own time. The explosion of the Krakatoa volcano in 1883 destroyed the entire habitat on Pulau Rakata, an island in the Sunda strait between Java and Sumatra. But life quickly reestablished itself there, in great diversity. Or consider the Galapagos Islands volcanoes, which are at most 3 million years old, and which have repeatedly erupted even in relatively recent times. As a group, the islands provide the basis for the splendid but still manageable animal and plant world that allowed Charles Darwin to discover, a century and a half ago, the principle of evolution: Small initial populations that spread into new habitats and can keep separate from one another are the raw material for evolution and a constant source of innovation.

Spoonbill

It makes more sense to call catastrophes in geological history "fauna turning points" or "flora turning points" when they have led to significant new beginnings. They do not represent decimations or extinctions of many species, but rather provide the crucial impetus for new developments. Imagine that the earth had never suffered catastrophes: What would have happened to life on the planet? Trying to answer this question is like trying to conceive humanity without crises, migrations, or disasters. Under unchanging, peaceful conditions, humans probably would never have become humans as we know them, but instead would have remained self-sufficient primates living in the forests, eating bitter leaves and sweet fruits.

However, life would never have reached even this stage had it never had to cope with overwhelming changes. We have to assume that the oceans would have remained filled with the simplest life-

Animal	Maximum Number of Offspring
Lions	Young per litter: 3
Brown rats	Young per litter: 9
Wild rabbits	Young per litter: 15
Mosquitoes	Eggs per laying: 150
Leatherback turtles	Eggs per laying: 1,000
Frogs	Eggs per laying: 12,000
Carp	Eggs per laying: 300,000
Honeybees (queen)	Eggs per day: 1,600
Termites (queen)	Eggs per day: 86,000

Life Insurance
For many animals, having large litters increases their chances of handing on their genes. Others ensure their reproduction by carefully protecting a few offspring.

forms, ones similar to bacteria. The development of higher organisms would never have occurred – because there would have been no reason for such development. In this sense, evolution needed catastrophes. The mutability of the earth – drifting continents and restless volcanoes, changes in seas and the alternation of warm, tropical periods and ice ages – provides the great external impetus for advances in evolution. However, this development is not at all accidental in the strict sense of the word. Organisms can change only on the basis of existing organisms, and cannot accidentally bring forth something wholly new.

The secret of the success of birds and mammals, after the extinction of the dinosaurs, lay in their inner organization, in their highly efficient metabolism. The latter was far more lavish than that of the dinosaurs, as we can tell by comparison with still-extant reptiles. It transforms five to ten times as much energy per gram of body weight, but in return, it enables birds and mammals to withstand changes in weather and variations in temperature between day and night. At night, when the saurians had to remain immobile – because it had grown too cold for them – the nimble little mammals could romp about. When winter came to large areas and reptiles went into hibernation, birds could fly away to warmer climes.

Pelican

Humans' opportunity came with the Ice Age. Its alternation of cold phases in which the ice advanced, and warm phases in which the ice retreated, imposed an unprecedented dynamic on the earth. Humans – primates that came from dwindling forest areas – were able to conquer the savannas produced by the alternation of arid and rainy periods, together with the wealth of large mammals that lived on them. The improvement in the quality and quantity of the meat humans ate was accompanied by an increase in the size of their brains, which ultimately exceeded that of all other animals.

Our own example shows that evolution follows no plan, and that it is at the mercy of the vicissitudes of geological and historical development. Yet pure chance does not rule. Life makes its own path. Through the catastrophes, it grew beyond itself, so to speak, and learned to free itself more and more from external conditions. This was and is the best life insurance. For the life-forms that survived all the catastrophes were the ones that had a wide range of options and could spread over the earth. Humans thus had – thanks to their expansion all over the planet, their highly differentiated life-forms, and their exceptional independence from immediate environmental conditions – the best chance of survival. That is what evolution teaches us, in opposition to all the doomsayers who predict that humanity has no future.

However, other life-forms pay a high price for our success. Many other species are in decline because humankind is spreading, and especially because around the world it is altering the environment to suit itself. The conditions of life have never before changed as rapidly as they have in the last 10,000 years and as they are still changing today. As a result, the relationship between species that are dying out and ones that are newly emerging has drastically changed. Humans have subdued the earth, as if they had set out to fulfill the biblical injunction (given in Genesis 1:28) only by extermination.

Avocet

Many scientific observers think the rate at which species are now dying out is several magnitudes greater than the rate at which new species are emerging. They base their opinion on the number of species that can be proven to have died out, chiefly as a result of human activity over the past century. Since the year 1600, the rate of the decrease in birds and mammals has increased at least fourfold. It must be correspondingly higher in groups of highly differentiated life-forms, and especially among insects, with their millions of species. They are far more closely connected with their specific habitats than are warm-blooded birds and mammals, and in some cases their habitats have shrunk by as much as 50 percent.

But there is still little sound scientific evidence for this. Calculating relative numbers that depend on the state of our knowledge about the overall number of species (which is frequently revised) is less important than assessing the consequences of this loss of species. Some scientists think these consequences are enormous, and may even lead the earth's ecosystem – including humans – to stop functioning. Other scientists maintain that all the previous losses in species neither endangered the overall ecosystem nor significantly altered the natural economy on the regional or local levels.

However, the natural economy is existentially important. This is no longer a question subject to debate; it is an established truth. We have long known that our existence depends on nature, and that despite all our technical achievements we still need good water, clean air, and healthful food. No one would really want nature to be reduced to a dozen kinds of food plants, along with cattle, pigs, sheep, and a couple of ornamental trees in glass-covered parks, with no more wild animals. Conservation, the maintenance of biodiversity and environments worth living in, has rightly become the central goal for the third millennium. The massive increase in the human population threatens to overwhelm and destroy biodiversity.

Yet in practice there will be many unanticipated obstacles to achieving this goal. Many people are unwilling to accept any restrictions for the sake of biodiversity. The basic problem often seems to be a misunderstanding that emerged in the nineteenth century as a result technological development along with inhospitable, dreary factories and the dirt, odors, and poisons that accompanied them. Wild areas were seen as the opposite, and a call went out for a "return to nature." In nature, according to this view, everything was beautiful, good, well-ordered, and healthy. It was humanity that was laying waste, disfiguring, poisoning, and destroying.

Rhinoceros hornbill

The notion of the economy of nature, inspired by the young, barely recognized science of ecology, has quickly taken hold in an almost religious way. In this economy, as in a well-conducted human economy, everything is supposed to be optimally regulated, adjusted, and functional. Every life-form has its place and its role in this economy. If one is removed from it, the result is a gap, a defect; in short, a weak spot in the system. This view accords very well with a notion that up to this point has played a central role in the observation of the living world of nature: the "balance of nature." According to this widespread concept, this balance has evolved over millions of years and must not be disturbed.

However, this idea is much more applicable to human beings and to their goals and requirements for life than to wild, undomesticated nature. In the human organism, all changes brought about by growth, development, and aging must in fact be kept in equilibrium by an "inner balance," homeostasis. This is achieved by many and diverse regulatory processes in the body. If they cease to function, we get sick. If the disturbance is too great, we die. These inner requirements are transferred to "nature" and its alleged balance. This transfer is facilitated by another concept drawn from modern ecology: that of the "ecosystem." From this concept it is concluded that nature is composed of ecosystems in the same way that an organism is composed of organs. The ecosystem thus quickly became a "superorganism." Seen in this way, it could "break down," "be destroyed," or be subjected to pollution and many other woes that can befall a living being.

Yet ecosystems are fundamentally different from organisms; they have no natural limits that separate them from the external world. Their functions have no central control like the one anchored in the genetic inheritance of living beings, and consequently they cannot reproduce themselves. Ecosystems are nothing other than research models. As such, they are suitable for answering certain questions, but in reality ecosystems do not exist, and in nature imbalances are far more common than balances. Whenever nature remains stable for a time, its efficiency decreases. Its diversity often decreases at the same time, and developmental processes lose their impetus. So we arrive once again at the question of whether life would have developed at all without disturbances, protected within an equilibrium.

Swift

Evolution, the splendid process of becoming, is dependent on disturbances, on imbalances that repeatedly appear: They are the source of its power. Had it remained in balance, it would have come to a halt. The romantic images of balance, harmony, and cycles are misleading. The air we breathe, with its oxygen, was not there from the beginning; it emerged from the massive overproduction of green plants over hundreds of millions of years – it is, in fact, their by-product, just as coal and petroleum are the massive surpluses of remote eras. The Alps and other great limestone mountain ranges are the by-products of living creatures that left behind enormous amounts of calcium.

From the outset, the relationship between plants and animals was, and still is, an imbalance. The same can be said of the relationship between predators and prey. Imbalances are the source of extinctions as well as of the emergence of new species. Without them no new species would have been able to establish themselves. Without variations in temperature or rainfall, habitats on land and in the sea would have been all alike. Only two monotonous habitats would remain, and in them only a few species could live because there would be no different niches in which specialized life-forms could find their place.

This does not mean, however, that nature, instead of being governed by harmony and order, would be governed by pure chaos. The dynamics of nature is neither order nor chaos, but rather a multilayered, constantly changing process intermediate between these poles; the process has its own inherent laws, but is not predetermined. Thus human beings, with their huge overall biomass, are a highly attractive space for the development of a wide variety of pathogens, parasites, and bloodsuckers. In the same way, the leaves, wood, and root masses of trees or the green plants of grasslands and tundras became the habitat for many life-forms. They have specialized in using a certain part of the habitat, and compete with each other. As we know from efforts to preserve plants and to fight carriers of disease, they undergo exceptionally rapid developments and changes: evolution in action.

Flamingo

Biodiversity is the result of change in the conditions and possibilities for life; it is life's response to the dynamics of nature. The processes of specialization and divergence are what allow diversity to emerge. This is shown in a particularly clear way by insects. In their world, a genus contains many, sometimes thousands of different species that look so much alike that only specialists can tell them apart. But insects' strategy is to divide up among themselves the possibilities for life, down to the last detail.

On single hectares in the rain forests of the central Amazon and of Borneo, 400 to 500 different species of trees have been found. Practically every tree belongs to a different species. In Venezuela, a middle-sized tropical country, there are more than 2,400 different species of trees; in the Upper Amazon, there are 180 different frog species per square kilometer. Indeed, more different species of ants may live on a large tropical tree than in all of England. Biodiversity is at its greatest in the tropics.

However, it is not the particularly propitious living conditions in the tropics that produce such abundance, but rather the opposite: It is chiefly lack that produces diversity. For where resources are scarce, specialization is worthwhile. And when habitats become islands as a result of changes in climate, different species often emerge alongside one another. For example, during the Ice Age tropical areas shrank and, as a result, were separated from one another, but when the climate grew warmer again they blended into each other once more. These processes require time. They proceeded slowly enough that the species were able to diversify themselves. In addition, their populations remained small enough that adaptive changes in their genetic heritage could rapidly establish themselves. Large populations are too sluggish for rapid evolution. Hence the tropics developed a habitat that has far more species than any other on earth.

If we examine more closely the process that has produced diversity of life, we inevitably encounter a riddle, the riddle of "species." Why does life divide into such a plenitude of different species that live separately from each other? So-called good species do not cross with other species in the neighborhood: They keep to themselves.

Merganser

All humans on Earth represent a single species because no human group or race is biologically cut off from the others. By definition, all individual life-forms that can mate with each other, and produce offspring that are able to do the same without restriction, belong to one and the same species. Conversely, if a barrier to such cross-breeding develops, then we are confronted with different species. Thus all breeds of dogs, no matter how different they might look, remain dogs, and are not different species (and cannot be bred into a new species). However, the many small birds that come to our bird feeders in the winter, and those that we can distinguish more or less well, belong to different species.

The situation is different in the case of plants. Many species hybridize fairly easily, especially with closely related species. Some food plants are descended from such hybridizations; for instance, corn was produced by crossing wild maize with teosinte.

Even if we do not examine the question so closely, there are good grounds for thinking that the "discovery of sexuality" by earlier life-forms led to the development of species. When the genetic heritage of an individual is brought together (crossed) with that of another in a very precise way, new combinations of genetic material are produced much more quickly than by random mutations. Sexuality is a good way to test the fitness of these new combinations.

In the early stages of the evolution of life, reproduction was asexual, like that of amoebae and other protozoans. The emergence of sexual reproduction – that is, reproduction that involved the intersection of the genetic heritages of two different individuals – vastly increased the speed of evolution. But the combination of very different genetic heritages is much more likely to result in offspring that are not viable. Species restrictions function to exclude most of these unviable outcomes. This may be what enabled more complicated, larger life-forms with internal organs or different developmental stages – egg-caterpillar-pupa-butterfly, for instance – to emerge.

Crossbill

Something else is also expressed in species: the almost infinite diversity of information regarding the environment. It is fixed in the genetic material and can be handed on to future generations. In this respect, species are information carriers and not mere forms of living creatures that have existed for a couple of hundred thousand or millions of years. They bear within themselves evolution's whole treasury of knowledge. This knowledge provides one of the most important arguments for preserving biodiversity.

Suppose we examine a small beetle that has just become extinct because its habitat in the tropical rain forest has been burned down, and which closely resembles countless other beetles. If we see it as an information carrier, it appears in a quite different light; it no longer seems unprepossessing and uninteresting, for what is at stake is the information acquired by evolution and stored in it. Perhaps this beetle has a gene that allows it – and it alone – to break down poisonous materials in plants and so make them harmless. Perhaps it has discovered how to derive from the poisonous substances in its food a chemical highly effective against certain bacteria or fungi. No one can say how the beetle might have been useful, if it still existed. The very fact that it survived for a long time makes it something special – with regard to the 99 percent of all species that died out before it.

Therefore research on biodiversity is valuable. We should determine as quickly as possible how many species there are that still live with us on the planet Earth. Then we would be better able to assess the losses.

Studies on the independence of species populations with regard to the surface area of their habitats are currently the only foundation for estimates of the number of species lost. We have more information only on birds and mammals that have long been known. According to biologist Andrew P. Dobson, since the year 1600, 113 species of birds and 83 species of mammals have become extinct worldwide. That might not seem like many when compared with a total population of 9,950 bird species and 4,630 mammal species. But the past four centuries represent only the final phase of the great extinction caused by humans, which began toward the end of the last Ice Age. Since then, about 2,000 species of birds have become extinct.

The very smallest animals were first scientifically studied only in the last 150 to 200 years; most of them may still not be known at all. Here we do not have hard numbers, but only estimates based on probability. Accordingly, Dobson estimates that, at present, between 4,000 and 14,000 species are dying out every year (10 to 38 per day). Over a thousand years – a very brief moment on the geological time scale – this would mean a loss of more species than are currently known, or about the same number as we can realistically assume to exist on Earth. If this is so, then in our time, species extinctions are occurring that are absolutely comparable to those produced by the greatest catastrophes in the history of the planet.

Eagle

In my opinion, these extrapolations vastly overstate the case. The major reduction in the loss of species between the nineteenth and the twentieth centuries, in all categories of vertebrates and nearly worldwide, offers grounds for a more optimistic assessment. In addition, this reduction also puts in question the basis on which Dobson calculated his estimate. We simply know too little about the pattern of the distribution of species, especially in the tropics. The calculations regarding the relation between the disappearance of species and the destruction of habitat are plausible only when most of the species occur in isolated patches, mosaiclike. But if the conditions are essentially like those in Europe, a different picture emerges. In Europe, nature has been transformed into an agricultural landscape in which only a few tiny "wild" spots remain. Yet in this partial continent, humankind's long-term interventions have not harmed the diversity of species. Many areas of Europe are even richer in species than they were before the advent of humans.

The truth must lie somewhere in between: The European-optimistic scenario cannot be right, because nature in the tropics is too different from it; but the statistical-pessimistic scenario cannot be right, either, because it greatly underestimates the adaptability and flexibility of most species.

All the same, concern is justified, and it is reasonable to take every feasible step to preserve the earth's biodiversity. Of this splendid abundance we currently know only about 1.75 million species. Unless we are absolutely forced to do so, we should not give up even the life-forms – sometimes magnificent, sometimes unappealing – that have emerged in the process of evolution and that have endured down into our own time. They are the capital of life for the future and the basis of its dynamics.

Yellowhammer

See also plates:

A Changing World
Page 24
Humans, a Career
Page 26

Hot Spots of Biodiversity
Page 32

Sources for text in margins:

1 M. Reaka-Kudla, D. E. Wilson, E. O. Wilson, eds.: Biodiversity II, 1997
2 Herder Lexikon der Biologie, 1992

Part Two
Use and Protection

The only way to help nature is to free it
from its apparent opposite, independent
thinking.
Max Horkheimer

Bison Had to Give Way to Cattle

The Return of the Bison
Before the arrival of white settlers, about 60 million bison lived in North America. In 1889, the number of bison in the United States had fallen to 500. Today, they number more than 200,000. This increase is due in part to the more than 2,300 bison ranchers in the United States. Between 20,000 and 25,000 bison are slaughtered every year in the United States. A female bison is worth 900 to 1,200 dollars, whereas a cow is worth only about 700 dollars.

Wild Bovines and Domestic Bovines
In North America, there are 113 million cattle and only 200,000 bison – for a ratio of 1:565. Over 99 percent of the cattle in the world are domestic animals. All eleven species of wild bovines, including bison, constitute only 0.06 to 0.11 percent of the total number of bovines in the world.

1:565

Bison Population, in Millions

| Year | 1700 | 1800 | 1889 | 2000 |

65 60 55 50 45 40 35 30 25 20 15 10 5

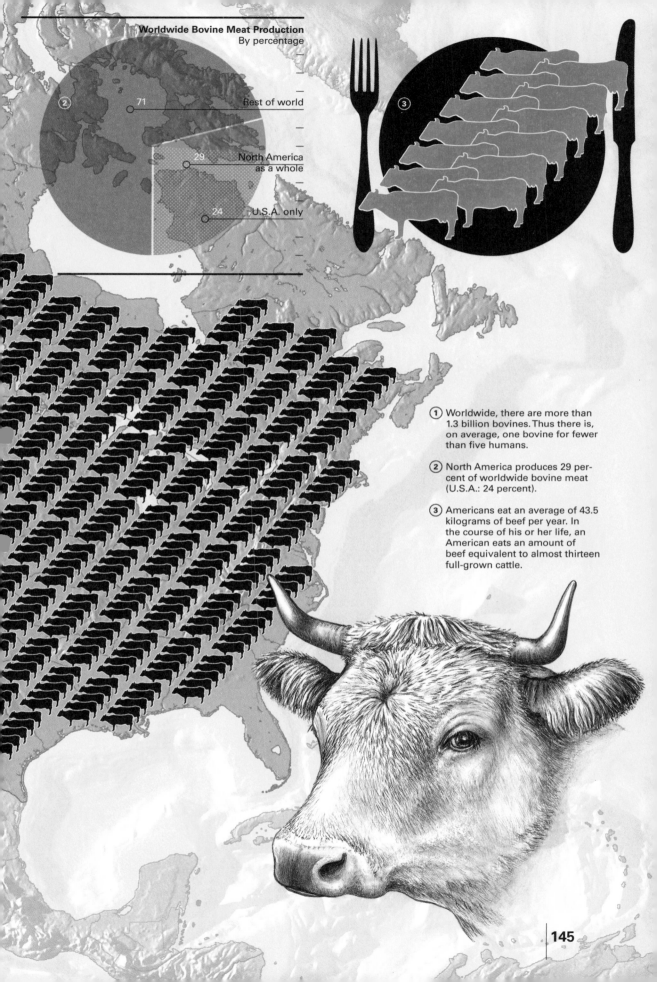

Worldwide Bovine Meat Production
By percentage

② 71   Rest of world

29   North America
as a whole

24   U.S.A. only

③

① Worldwide, there are more than
1.3 billion bovines. Thus there is,
on average, one bovine for fewer
than five humans.

② North America produces 29 percent of worldwide bovine meat
(U.S.A.: 24 percent).

③ Americans eat an average of 43.5
kilograms of beef per year. In
the course of his or her life, an
American eats an amount of
beef equivalent to almost thirteen
full-grown cattle.

145

The World's Three Food Sources

Only a tiny fraction of the diversity of plant life is eaten
Of the thousands of edible plants that grow in the wild, humans use astonishingly few. Over 80 percent of the human food supply comes from only a dozen cultivated plants. The nutritional foundation for most people consists of three plants: wheat, which was one of the first plants to be cultivated, about 10,000 years ago in the Middle East; rice, which was first developed in China; and corn (maize), which originally came from Central America. The cultivated varieties of these plants are highly genetically diverse: 3,000 kinds of wheat, 5,000 kinds of rice, and 6,000 kinds of corn. However, most farmers in industrial countries sow only a few high-yield varieties.

So many humans are dependent on wheat, rice, and corn
Wheat is the most important plant in the world. For 3.1 billion people (54 percent of the world's population), it represents the basic element of the diet (see chart). Next comes rice, which is the main foodstuff for nearly 2 billion people (34 percent). Corn is the staple of the diet for 0.7 billion people (12 percent). Afterward come potatoes, barley, manioc, sweet potatoes, and soybeans.

Today, on average, 32.5 percent of a human's diet consists of animal products, chiefly milk.

About 67.5 percent of a human's food consists of plants. This corresponds to about 273 kilograms per year.

3 major food sources
corn, wheat, rice

15–20 food plants
of economic significance

200 cultivated species
of food plants

3,000 species
possible food plants

270,000 species
known higher plants

Corn

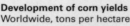
Corn kernel

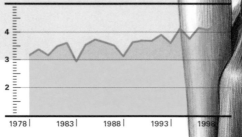

Development of corn yields
Worldwide, tons per hectare

4
3
2

1978   1983   1988   1993   1998

Development of yields, 1978–1998
By using better techniques, the yield per hectare has been significantly increased (see chart). Since the early 1970s, worldwide food production has risen by 80 percent. Within the next 25 years, yields must increase by 50 percent if the world's population is to be fed.

How much a kilogram of rice costs in U.S. dollars (1999)
In wealthy industrial countries such as the United States, food is relatively inexpensive in relation to income, whereas in developing countries many families have to spend nearly all their money on food.

U.S.A.	Brazil	Nigeria	India	China
0.9 – 1.7	**0.58 – 0.68**	**0.55**	**0.19 – 0.94**	**0.16 – 0.2**

Wheat

Rice

Wheat kernel

Rice kernel

A human pictogram = one billion people

Development of Wheat Yields
Worldwide, tons per hectare

Development of Rice Yields
Worldwide, tons per hectare

4

3

2

1978   1983   1988   1993   1998

4

3

2

1978   1983   1988   1993   1998

The Exploitation of the Seas

Overfishing: a Global Problem
For centuries, the numbers of fish caught were able to increase steadily. Toward the end of the twentieth century, however, the threshold for sustainable long-term use was crossed. Some fishing grounds are entirely exhausted. In whole regions, this has caused people in the fishing industry to lose their jobs. This uneconomic overexploitation is subsidized by many countries. The European Union countries alone subsidize their fishing fleets to the tune of 1.5 billion U.S. dollars a year. 20 percent of the cost of each kilogram of fish caught has already been paid by taxpayers. Only in the 1990s did the international fishing industry begin to develop concepts for long-term, sustainable fisheries.

Atlantic herring

Alaska pollock

Anchoveta

Fish farming is becoming increasingly important. 180 million young fish from fish farms are released daily into the seas and aquaculture operations.

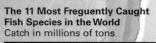

180,000,000

Fish is the most important natural food source for humans. On average, each individual consumes 15.7 kilograms per year.

The 10 Most Important Fishing Grounds in the World
Catch in millions of tons

Country	Catch
China	14.222
Peru	9.515
Chile	6.692
Japan	5.964
U.S.A.	5.000
Russia	4.675
Indonesia	3.729
India	3.491
Thailand	3.138
Norway	2.638

The 11 Most Frequently Caught Fish Species in the World
Catch in millions of tons

Species	Catch
Anchoveta	8.863
Alaska pollock	4.533
Chilean jack mackerel	4.378
Atlantic herring	2.330
Chub mackerel	2.167
Capelin	1.527
South American pilchard	1.493
Skipjack tuna	1.479
Atlantic cod	1.329
Largehead hairtail	1.275
Japanese anchovy	1.254

Changes in the Catch and in Fish Farming Worldwide
In millions of tons, 1950–1997

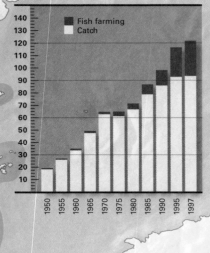

■ Fish farming
□ Catch

1950 1955 1960 1965 1970 1975 1980 1985 1990 1995 1997

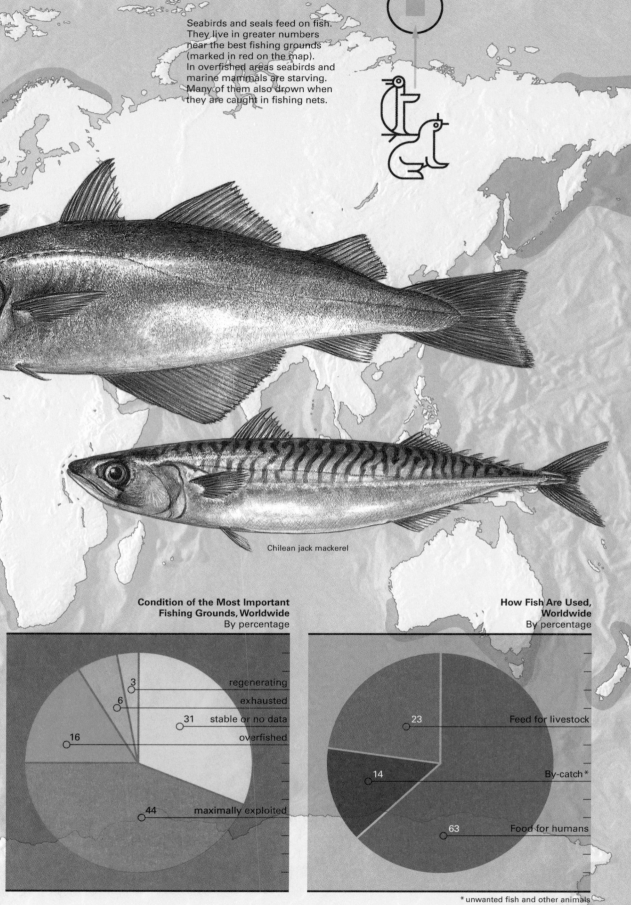

Seabirds and seals feed on fish.
They live in greater numbers
near the best fishing grounds
(marked in red on the map).
In overfished areas seabirds and
marine mammals are starving.
Many of them also drown when
they are caught in fishing nets.

Chilean jack mackerel

**Condition of the Most Important
Fishing Grounds, Worldwide**
By percentage

3 — regenerating
6 — exhausted
31 — stable or no data
16 — overfished
44 — maximally exploited

**How Fish Are Used,
Worldwide**
By percentage

23 — Feed for livestock
14 — By-catch *
63 — Food for humans

* unwanted fish and other animals
that are caught and thrown
back into the sea (usually dead)

Forests as an Economic Factor

Forests and Humans
More than 3.4 billion hectares of forest cover our planet – 26.6 percent of its land surface. The word "forest" includes various kinds of arboreal vegetation: tropical rain forests, northern conifer forests, and many others. Most are used and altered by humans. Worldwide, in 1996 nearly 1.5 billion cubic meters of logs were milled and more than 1.8 billion cubic meters were used as firewood. During the first half of the 1990s, forested areas decreased by 0.3% (over 56 million hectares) worldwide. Although they are increasing in industrial countries, they are shrinking in tropical areas. Rain forests are being cleared to make way for plantations producing palm oil, latex, and other products.

Endangered Mangroves
Particularly threatened are mangroves in tropical coastal areas. Large mangrove forests were logged off in the 1980s and 1990s in order to construct shrimp farms (in Thailand, Ecuador, Indonesia, China, and India, for example). As a result, the breeding grounds for many species of fish that grew up among the mangrove roots were destroyed. Clearing mangroves also endangers humans because mangroves protect and stabilize coastal areas. When they are removed, tidal waves can do greater damage.

The 5 Countries with the Largest Wood Plantations
Area cultivated, in millions of hectares)
1. China (33.8)
2. India (14.6)
3. Indonesia (6.1)
4. Brazil (4.9)
5. Vietnam (1.5)

Estimated area of tropical rain forests (marked in black) in South America, at the end of the last Ice Age.

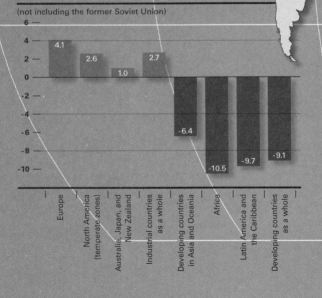

Growth and Loss between 1980 and 1995
Changes in forested area, by percentage

(not including the former Soviet Union)

Region	Percentage
Europe	4.1
North America (temperate zones)	2.6
Australia, Japan, and New Zealand	1.0
Industrial countries as a whole	2.7
Developing countries in Asia and Oceania	-6.4
Africa	-10.5
Latin America and the Caribbean	-9.7
Developing countries as a whole	-9.1

The 7 Countries with the Largest Forested Areas
(Share of worldwide forested area, by percentage)
1. Russian Federation (22.1)
2. Brazil (15.9)
3. Canada (7.1)
4. U.S.A. (6.2)
5. China (3.8)
6. Indonesia (3.2)
7. Democratic Republic of Congo (3.1)

**Worldwide, There Are 6,000 Square
Meters of Forest per Person**
This area corresponds to four-
fifths of a soccer field. There are
about 500 trees per person
(average based on the tropical
rain forest and central European
deciduous forest). In 1995,
humans used 3.3 billion cubic
meters of wood, or 0.6 cubic meters per capita.

**The Five Nations that Export
the Most Wood Products**
(Exports in millions of U.S.
dollars, 1996)
1 Canada (25.3)
2 U.S.A. (16.9)
3 Sweden (11.0)
4 Finland (10.3)
5 Germany (9.4)

Trees Are Habitats
Inventories show that in Great
Britain there are 465 species
of arthropods (insects, spiders,
millipedes) that are specially
adapted to oak trees. In Panama
alone there are 1,200 species
of beetles that are specially
adapted to tropical trees of the
species *Luehea seemanii*.

One Tree, Many Products

The Coconut Palm
Humans use many plants – hemp and bamboo, for example – in different ways. Yet few plants offer as many possibilities as the coconut palm, which grows throughout the tropics. It is one of about 2,800 palm species. The nut of the Seychelles coconut palm (*Lodoicea seychellarum*) is the largest seed in the plant world: 60 centimeters long and 22.5 kilograms in weight. Since the coconut palm makes few demands and since every part of the tree can be used, it has become a very important economic factor, especially in tropical regions. In 1997, 158,551 tons of coconuts were exported worldwide. In 1997, 2,006,240 tons of coconut oil were exported worldwide. The market value of coconut oil amounted to 1,288,132,000 U.S. dollars.

Shell fiber
rope
mats
brushes
upholstery
sandals

Copra
(grated and dried meat of the coconut)
coconut oil
fat
cosmetics
detergent
soap

Coconut milk
beverage

Residue from Oil Pressing
livestock feed

Shell
burned for heat
containers
veneers

Predicted Development of Coconut Oil Production
In millions of tons

Philippines
Rest of the world
Indonesia
Europe/U.S.A./Japan

1,2
1,0
0,8
0,6
0,4
0,2

| 1968-72 | 1978-82 | 1988-92 | 1998-2002 |

Share of Worldwide Coconut Plantation Area
By percentage, 1996

Philippines — 29.3
Indonesia — 24.1
India — 16.7
Rest of the world — 29.9

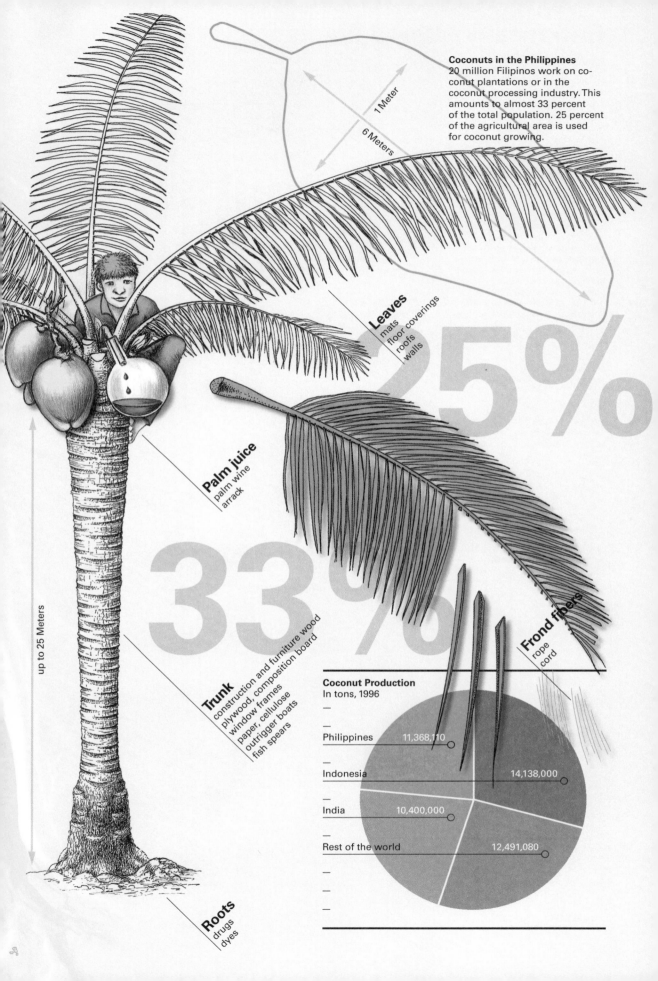

Coconuts in the Philippines
20 million Filipinos work on coconut plantations or in the coconut processing industry. This amounts to almost 33 percent of the total population. 25 percent of the agricultural area is used for coconut growing.

1 Meter

6 Meters

25%

33%

Leaves
mats
floor coverings
roofs
walls

Palm juice
palm wine
arrack

Frond fibers
rope
cord

Trunk
construction and furniture wood
plywood, composition board
window frames
paper, cellulose
outrigger boats
fish spears

up to 25 Meters

Roots
drugs
dyes

Coconut Production
In tons, 1996

Philippines	11,368,110
Indonesia	14,138,000
India	10,400,000
Rest of the world	12,491,080

Genetically Altered Plants

Biotechnology awakens hopes and fears

For thousands of years, humans have been altering cultivated and ornamental plants through breeding. With triticale, created by crossing rye and wheat, breeding was even able to cross species barriers. In order to produce more mutations (the only way to improve yield or quality), breeders are using increasingly refined methods. In the last third of the twentieth century, plant breeding – like animal breeding – entered a new stage of development. It became possible to isolate specific desired qualities in one plant's genetic material and transfer them to another plant. Genetic material can now be transferred to plants not only from plants of a different species, but even from animals or bacteria. This biological revolution, which allows humans to manipulate the genetic code of life more profoundly than ever before, awakens great hopes and also great fears. While at the beginning of the twenty-first century genetically altered plants are growing on an increasing number of farms all over the world, we are still far from achieving a consensus regarding this new means of food production.

Acceptance of Agricultural Genetic Technology in Europe
By percentage, 1996

- accept
- reject
- undecided

	accept	reject	undecided
Europe as a whole	59.5	28.8	11.8
Germany	54.9	32.8	12.2
Greece	59	24.5	16.5
France	65	27.2	7.7
Great Britain	64	21.8	14.3
Sweden	59.9	34.8	5.3

Acceptance of Medical Genetic Technology in Europe
by Percentage, 1996

- accept
- reject
- undecided

	accept	reject	undecided
Europe as a whole	69.1	19	11.9
Germany	65.9	22.4	11.6
Greece	64.6	16.6	18.8
France	74.1	17.6	8.4
Great Britain	73.5	13	13.5
Sweden	78.4	15.5	6.2

50 percent of all seed producers for grains use genetic material from wild forms, which they obtain from gene banks.

The Most Important Genetically Altered Plants, Worldwide
By percentage of area under cultivation, 1998

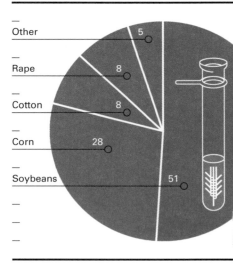

- Other 5
- Rape 8
- Cotton 8
- Corn 28
- Soybeans 51

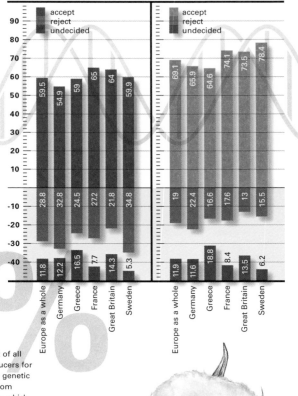

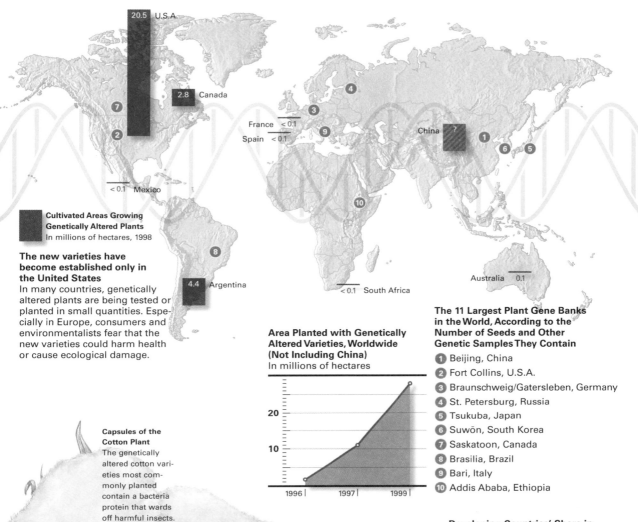

20.5 U.S.A.

2.8 Canada

France < 0.1
Spain < 0.1

China 7

< 0.1 Mexico

**Cultivated Areas Growing
Genetically Altered Plants**
In millions of hectares, 1998

**The new varieties have
become established only in
the United States**
In many countries, genetically
altered plants are being tested or
planted in small quantities. Especially in Europe, consumers and
environmentalists fear that the
new varieties could harm health
or cause ecological damage.

8

4.4 Argentina

Australia 0.1

< 0.1 South Africa

**Capsules of the
Cotton Plant**
The genetically
altered cotton vari-
eties most com-
monly planted
contain a bacteria
protein that wards
off harmful insects.

**Area Planted with Genetically
Altered Varieties, Worldwide
(Not Including China)**
In millions of hectares

20

10

1996   1997   1999

**The 11 Largest Plant Gene Banks
in the World, According to the
Number of Seeds and Other
Genetic Samples They Contain**

1 Beijing, China
2 Fort Collins, U.S.A.
3 Braunschweig/Gatersleben, Germany
4 St. Petersburg, Russia
5 Tsukuba, Japan
6 Suwŏn, South Korea
7 Saskatoon, Canada
8 Brasilia, Brazil
9 Bari, Italy
10 Addis Ababa, Ethiopia

**Developing Countries' Share in
Genetic Research in the Area of Agriculture**
By percentage, 1998

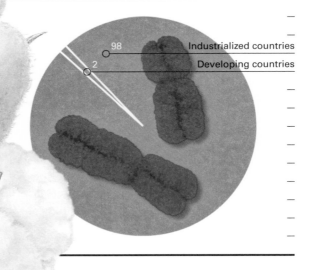

98 —— Industrialized countries

2 —— Developing countries

155

The Green Pharmacy

Plants and Animals Save Human Lives

At the beginning of the twenty-first century, trade in medicines made from plants is supposed to reach an estimated $500 billion. At the end of the 1990s, a quarter of drugs prescribed in the United States were already derived from plants. In 1996, about half the drugs commonly used in Germany contained ingredients derived from plants. Biologists and pharmaceutical researchers are studying wild plants and animals in search of substances that can be extracted or synthesized to make new drugs. At the same time, traditional herbal medicine is booming in Asia and Europe. 90 percent of all medicinal herbs are not cultivated but collected in the wild, and in many cases this endangers the existence of whole species.

From Natural Material to Drug

The path from the wilderness to the pharmacy is a long one. Plant extracts are made from as many as 300 pure substances, all of which have to be researched. The chart to the right shows that thousands of natural substances are analyzed, altered, and tested in order to develop a marketable drug. This process takes from ten to fifteen years.

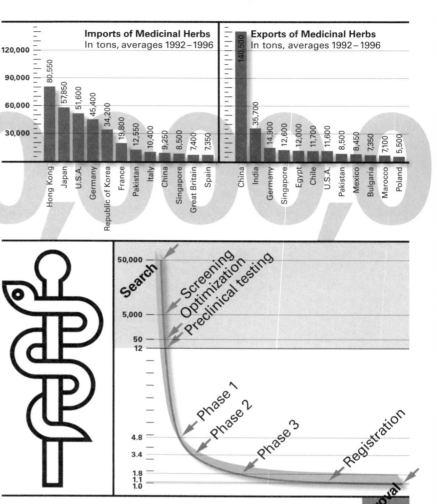

Imports of Medicinal Herbs
In tons, averages 1992–1996

Country	Tons
Hong Kong	80,550
Japan	57,850
U.S.A.	51,600
Germany	45,400
Republic of Korea	34,200
France	19,800
Pakistan	12,550
Italy	10,400
China	9,250
Singapore	8,500
Great Britain	7,400
Spain	7,350

Exports of Medicinal Herbs
In tons, averages 1992–1996

Country	Tons
China	140,500
India	35,700
Germany	14,900
Singapore	12,600
Egypt	12,000
Chile	11,700
U.S.A.	11,600
Pakistan	8,500
Mexico	8,450
Bulgaria	7,350
Marocco	7,100
Poland	5,500

Search — Screening — Optimization — Preclinical testing — Phase 1 — Phase 2 — Phase 3 — Registration — Approval

50,000 / 5,000 / 50 / 12 / 4.8 / 3.4 / 1.8 / 1.1 / 1.0

In Search of Healing Powers

Universities, research institutes, and pharmaceutical firms have sought usable substances in plants and animals throughout the world – and even in the seas. At the end of the 1990s, for instance, materials taken from algae, corals, sea snails, and sharks were being tested as drugs. The map shows a few examples of this kind of bioprospecting in which institutions and companies from the United States are participating.

Two Plant Substances of Medicinal Significance

– Virblastine and vincristine, used in fighting leukemia in children and Hodgkin's lymphoma, are derived from Madagascar periwinkle (*Catharanthus roseus*).
– Taxol, used in the treatment of ovarian cancer and metastasizing breast cancer, is derived from the Pacific yew tree (*Taxus brevifolia*).

Two Animal Substances of Medicinal Significance

– A poisonous alkaloid, taken from the saliva of a small frog (*Epipedobates tricolor*) that lives in the Andes, is 200 times more effective than morphine in relieving pain.
– Ancrod, which dissolves blood clots, is used in treating heart attack patients. The enzyme is derived from the poison of the Malayan pit viper.

U.S.A., Yellowstone National Park: National park administration and a biotechnology company

Mexico: A biotechnology company

Ecuador, Amazon area: A pharmaceutical company seeking information from original inhabitants with medical knowledge

Argentina, Chile: University of Arizona, Louisiana State University, and a pharmaceutical firm

Costa Rica: The Costa Rican Instituto National de Biodiversidad, Cornell University, and two pharmaceutical firms

Surinam: Missouri Botanical Garden and a pharmaceutical firm, sponsored by governmental scientific and developmental aid organizations

Peru: Washington University and a pharmaceutical firm

China, Malaysia: National Cancer Institute

Indonesia: A biotechnology company

Cameroon: The Smithsonian Institution, Walter Reed Army Institute of Research, and two pharmaceutical firms

Tanzania: The Tanzanian Institute for Traditional Medicine and a pharmaceutical firm are seeking information from villagers with medical knowledge

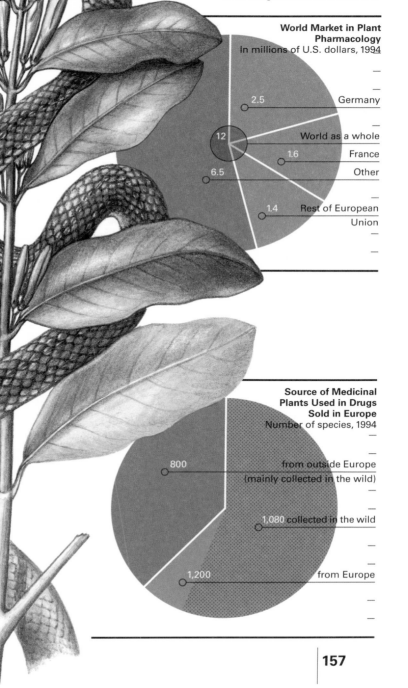

Asclepius's Snake

The Asclepius adder appears on the caduceus, the insignia of physicians and pharmacists. The symbol comes from ancient columns with figures showing Asclepius (Latin *Aesculapius*), the Greek god of medicine. He supports himself on a staff, around which a snake winds. According to the myth, the snake is supposed to have called his attention to plants with curative powers.

Virtual Encounter

The Asclepius adder lives in Europe and Asia Minor. The plant *Catharanthus roseus* grows in Madagascar and contains a substance used to fight leukemia in children.

World Market in Plant Pharmacology
In millions of U.S. dollars, 1994

- 2.5 — Germany
- 12 — World as a whole
- 1.6 — France
- 6.5 — Other
- 1.4 — Rest of European Union

Source of Medicinal Plants Used in Drugs Sold in Europe
Number of species, 1994

- 800 — from outside Europe (mainly collected in the wild)
- 1,080 — collected in the wild
- 1,200 — from Europe

Nature as a Productive Force

Wild Areas Create Value
Humans use only a small fraction of the earth's biomass – mainly cultivated plants and economically useful animals. However, agriculture and industry also rely on wild plants and animals. Soil organisms help make fields fertile, insects pollinate fruit crops, and forestry and fishing depend on the output of forests and seas. Beautiful landscapes are the raw materials of the tourist industry. Economists trying to assess the economic value of wild areas have arrived at astounding estimates. A hectare of tideland on the coast of the United States is worth 72,000 U.S. dollars a year in protection against flooding and as a breeding ground for fish.

85	Forest
200	Pharmaceuticals
2,000	Tourism

Estimated World Market of 3 Branches of Industry That Are Dependent on Nature
In billions of U.S. dollars

Insects pollinate fruit and vegetables
In the United States, the value of pollination by honeybees (shown here on an apple blossom) is estimated to be 6.7 billion dollars per year.

172.5	Net Primary Production, in Billions of Tons per Year *
	Animal biomass: 2.3
1,850	Total biomass

The World's Biomass, in Billions of Tons
including humans, economically useful animals, and cultivated plants. Most of the biomass consists of plants; humans and their domestic animals constitute one-fourth of the animal biomass.

* including 6 percent cultivated plants. The net primary production is the annual increase in plant biomass (except water plants). It is partly consumed by humans, and partly remains standing as trees, or rots and becomes earth.

Wild Animals as Food

Fish and wild game constitute 20 percent of humans' protein intake. In rural areas of some developing countries, the proportion is significantly higher. In the Democratic Republic of Congo, it is 75 percent; in the Ucayali region of Peru, it is 85 percent. 40 percent of the meat consumption in Botswana and 20 percent in Nigeria consists of wild game and fish. Insects, with a raw protein content of up to 80 percent, are also eaten in many countries. In Mexico alone, 200 species are eaten (worldwide, 500 species). 1,600 tons of roasted caterpillars are sold on the South African market every year.

Wild Animals as Attractions

A few countries that earn money from nature tourism.

U.S.A.: Annual revenues from tourist scuba diving on the Florida reefs: 1.6 billion U.S. dollars.

Botswana: Annual revenues from nature tourism: 10 million U.S. dollars.

Rwanda: Annual revenues from tourism to view gorillas: 1 million U.S. dollars (beginning of the 1990s).

Tanzania: Annual revenues from nature tourism: 27 million U.S. dollars (1980s).

Kenya: Annual revenues from nature tourism: 400 million U.S. dollars (beginning of the 1990s).

Maldive Islands: Annual revenues from diving fees: 17 million U.S. dollars.

Australia: Annual revenues from tourist diving on the Great Barrier Reef: 1.5 billion U.S. dollars.

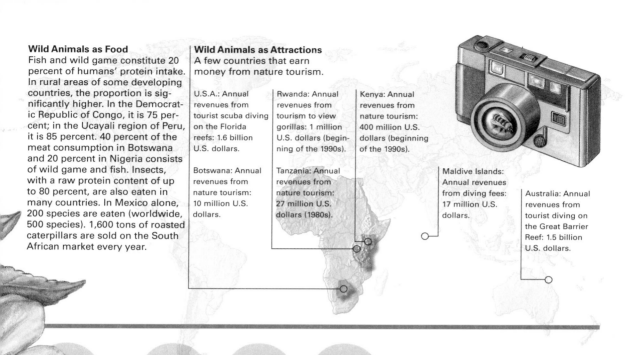

12	Medicinal plants
3	Other (meat, furs, live animals, etc.)
52	Fish
69	Wood

Worldwide Exports of All Products Made from Wild Plants and Animals
The total amount is almost twice that of the world's seven largest producers of computer chips.

136 Total

57.60	Rubber
27.00	Vegetables
22.50	Wild cocoa
199.67	Other edible fruits
391.02	Palm fruits

Annual Production per Hectare of Rain Forest in Peru
After three years, the ongoing profit from fruits and vegetables is already greater than the value of the wood of all the trees on the same hectare.

697.79 Total

Hunting blocks	930,000
Licenses, etc.	133,974
Contributions to conservation	1,256,700
Hunting fees	377,850
Trophy fees	199,850
Game fees	4,529,265

Revenues from Hunting Tourism in Tanzania
In U.S. dollars, fiscal year 1995–96. A large part of this revenue is invested in conservation.

Total 7,427,639

How Long It Takes for a Wild Animal Population to Double

In years. Wild animal populations grow with surprising speed if they are undisturbed by humans and there are no epidemics.

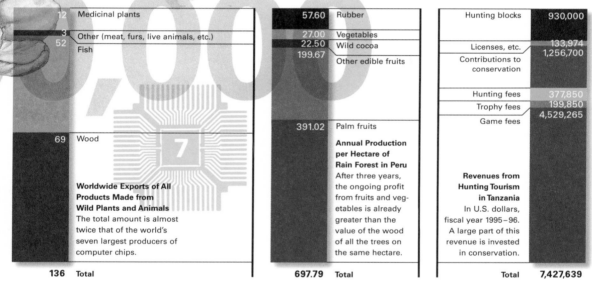

Calcite Production from Corals in the Australian Great Barrier Reef: 100 Million Tons Yearly

This corresponds to 17 times the weight of the Great Pyramid at Giza. The death of living creatures produces new material.

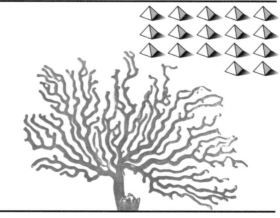

Human's Housemates

Domestic Animals and Ornamental Plants

For millions of families, animals and plants are part of their quality of life. In industrialized countries, the markets for plants, garden supplies, animal feeds, and veterinary medicine have developed into an important economic area.

Domestic pug dog. The origin of this breed is obscure, but it was probably first bred in China and then brought to Europe.

17,500,000,000

The worldwide gross economic value of all ornamental plants is between 16 billion and 19 billion U.S. dollars. This corresponds to the gross domestic product of whole countries (for example, Uruguay or Tunisia). Lawns and other garden plantings cover enormous areas.

11,400,000,000

Sales volume on the market for domestic animal care in Western Europe (European Union countries plus Norway and Switzerland) amounts to 11.4 billion U.S. dollars.

41,000,000

Domestic dogs in Western Europe

41 Million Dogs in Europe

In the form of the domesticated dog, the wolf has succeeded throughout the world. Dogs live on all continents, and in far greater numbers than wolves ever existed. Although wolves were almost completely exterminated in Western Europe more than a hundred years ago, at the beginning of the twenty-first century there are again between 3,000 and 3,500 wolves living in the wild in Western Europe.

261 million domestic animals (including fish in aquariums) are kept in European Union countries, Norway, and Switzerland.

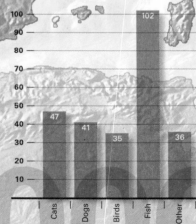

Domestic Animals in Western Europe
In millions, 1996

Animal	Millions
Cats	47
Dogs	41
Birds	35
Fish	102
Other	36

261,000,0

4,700,000

4.7 million tons of domestic animal food are produced annually in Western Europe. This is more than Switzerland's total production of cow milk.

350

On average, a Swiss citizen spends 350 U.S. dollars a year for cut flowers. That corresponds to nearly the per capita expense on body care products.

1.33

In 1999, 300 grams of a premium dog food cost the equivalent of 1.33 U.S. dollars in Germany. 300 grams of breakfast bacon for human consumption costs about the same.

Wolves are the ancestors of all dogs. Despite the changes resulting from many thousands of years of breeding by humans, dogs and wolves still belong to the same biological species and can interbreed

3,250

Wolves live in the wild in Western Europe

Imports of Cut Flowers and Potted Plants
World market share in millions of U.S. dollars

200	Israel
220	Denmark
250	Italy
540	Colombia
3,090	Holland
1,690	Other

Exports of Cut Flowers
World market share by percentage

1	Kenya
2	Spain
4	Israel
6	Italy
10	Colombia
59	Holland
18	Other

Sustainable Use: A New Method for Protecting Species

Guest Author: Prof. Klaus Töpfer, Executive
Director, United Nations Environmental
Programme (UNEP). A key to successful
protection of species lies in the concept of
sustainable use. The "management" of wild
animals can help preserve them.

The dramatic loss of animal species living in the wild over the
past few decades offers sufficient grounds for reflecting on the best
strategy for conservation. Although in the past most parts of the
planet went through phases in which wild animal populations decreased, the situation in developing countries is becoming acute:
The number of animal species that have recently been driven out
of their hereditary territories or completely wiped out is cause for
grave concern.

The two major threats to these populations are overhunting and
the large-scale loss of their natural habitats, the latter being the
more serious danger. The rapid increase in human population
means that a constantly increasing amount of land is needed for
farming and the raising of livestock – and is thus taken away from
wild animals. In Africa alone about two-thirds of all natural areas
have been sacrificed to farming and other uses.

The majority of the remaining habitat for animals living in the
wild is in developing countries. Yet retaining all or at least most of
these areas as wildlife reserves is not feasible because of the demand

for economic development. Population growth and poverty often leave no choice other than to transform wild areas into farms and pastures. Only a few of the areas that are still untouched are officially protected, and in many countries these reserves are too small to ensure the long-term survival of the fauna. This makes it urgent to protect wild animals outside reserves as well – precisely where they are particularly endangered.

In many places, sustainable use of wild animals has proven to be a useful tool both for preserving species and for serving the interests of the people of the country. Sustainable use merely skims off the natural surplus and maintains animal populations over the long term. The main goal is to stop the extermination of species. Toward this end, a system offering a country's people incentives to ensure that the long-term survival of wild animals in their region is set up.

Antelopes, buffalo, crocodiles, and birds all have a significant economic value. When it is possible to transform this value into cash – in the form of tourist revenues and hunting permits, for instance – people's attitude toward conservation usually changes as well. Their interest in preserving wild creatures and their habitat is awakened. In short, this prudent use of natural resources can be an effective way of reconciling conservation and economic development.

The traditional, "protective" approach to conservation (that is, without use) has its roots in the Western environmentalist movements of the past century. This approach, which is based on the principles of command and control, sought to protect nature from human abuse in two ways: by creating wildlife reserves and by prohibiting exploitation of species living in the wild. This may seem reasonable when seen from the point of view of someone living in a Northern Hemisphere country, but it gets nowhere in protecting nature in developing countries. Forbidding the use of wild animals amounts to declaring hunting that has been carried out for centuries – in order to sustain human life – to be a crime. The result is not merely to alienate humans from nature: Logically enough, they cease to regard animals as valuable resources, and see them only as pests and threats.

Gazelles are economically more important than cattle Scientists who did a study in Kenya discovered that the same amount of pasture land needed by one cow can feed eight gazelles. And the net profit from using fifteen species of wild animals was ten times greater than that from a comparable number of cattle. [1]

People living in rural areas have an entirely different idea of the animal world than do city dwellers, who do not need to share their habitat with large wild animals. It is the former who have to bear the costs resulting from conservation in the form of damage to their crops and loss of livestock. Wild animals compete for food with livestock herds, and some of them can even become a danger to humans. This explains why complete protection for wild animals is often unable to prevent poaching. On the contrary, sometimes it even leads to species of animals being totally wiped out.

If wild animals are not threatened by poaching, they are usually threatened as a result of the fact that they claim territory for themselves – valuable land. Farmers in poor regions of Asia, South America, and Africa, who often have to fight just to survive, have hardly any other choice than to get the most they can out of the land on which they live.

Wild animals compete with humans for the land, which is utilized not only for food production, but also for infrastructure and other uses. When wild animals are deprived of their economic value, they lose their significance as natural resources, and the transformation of their habitats is inevitable. Conservation conceived as merely "protective" and as not including use – a conception still favored by many conservationists – does not take this danger into account.

One possible solution consists of allowing the local population to use biological resources to support themselves. These resources include wild animals as well as other products of the bush and the forests. From an ecological point of view, cattle raising is a less attractive alternative; it has long been demonstrated that domestic animals such as cows and goats do extensive damage to ecosystems because their way of grazing is not in harmony with the composition of the soils. The original, indigenous species are much better adapted and consequently better suited to maintaining ecosystems.

What specific steps are required to achieve sustainable use? According to the Convention on Biological Diversity, sustainability means "the use of components of biological diversity in a way and at a rate that does not lead to the long-term decline of biological diversity, thereby maintaining its potential to meet the needs and aspirations of present and future generations." In many developing countries, the government owns, manages, and controls natural resources. So the first step toward sustainable use is decentralization – a transfer of the ownership or the right to use wild animals from the central government to the local population – so that making use of them becomes legal, and markets for wild animal products can be created.

The creation of such markets is an important step in transforming natural products into cash, making it possible for villages to get income from the use of wild animals. In order for this income to be fairly distributed, common institutions for managing wild animals have to be established.

Any kind of trade in wild animals has to be carefully overseen. It should be based on scientific knowledge of ecosystems in order to prevent overuse and other negative influences on nature. Such monitoring involves regular surveys of the habitat and animal populations, and the resulting census figures and assessments must be handled by independent scientists.

Committees and oversight mechanisms must enforce regulations and try to discover any abuses. In many cases, it has proven useful for villagers to elect game wardens from their own ranks; this increases the society's control over wild animals and the communities' sense that the resources actually belong to them. It is particularly important that village communities take an active role in all decisions and in management. This ensures that the program responds to the needs, culture, and wishes of the villagers as well. The central element of sustainable use creates positive incentives rather than threats and penalties.

In developing countries, many examples show that sustainable use of wild animals is at least as economically advantageous as agriculture. Often, such projects have strengthened the protection of biodiversity simply by preventing the transformation of natural habitats into agricultural land and by stabilizing the populations of the species used.

A Comparison of the Use of Wild Animals and Cattle-Raising in Botswana
(at 1991 prices)

Kind of use	Investment	Interest on capital
	(per hectare, in U.S. dollars)	(by percentage, over 10 yrs.)
Phototourism	15	28
Trophy hunting	0.7	38
Wild animal raising	32	7
Ostrich raising	5,569	19
Crocodile raising	11,229	11
Cattle raising	17	2

Economics and Wilderness
Hunting tourism in rural areas of Africa yields high profits for a minimal investment, and these profits benefit the local populations. [2]

The best known initiative based on incentives for use is probably CAMPFIRE (Communal Areas Management Program for Indigenous Resources) in Zimbabwe. It has existed since 1990 and currently covers an area of some 13,000 square kilometers. The central government transferred responsibility for the management of the populations of antelopes, gazelles, buffalo, zebras, and elephants in the project region to the communes. Measures were adopted to oversee use and to ensure a fair distribution of the proceeds among the government, the communal administration, and the villages. In many regions, communities now earn more from managing wild animals than from agriculture.

The rural population has a much more positive attitude toward conservation than it used to, poaching has greatly decreased, and animal populations are growing. Similar village projects – including raising butterflies, crocodiles, and sea turtles – have been set up in Botswana, Chile, Malawi, Mozambique, Namibia, Papua New Guinea, Peru, South Africa, and Zambia, although these have a considerably smaller scope.

In southern Africa, the use of wild animals has recently achieved particular success on lands in private ownership. In Namibia and Zimbabwe, about 75 percent of all livestock raising operations in semiarid areas have either added animals living in the wild or even shifted entirely to game farming.

An increasing number of landowners are selling their entire livestock herds in order to raise antelopes, gazelles, zebras, rhinoceroses, and elephants instead. Through the transformation of cattle farms into wild animal farms, the habitat for wild animals has increased by several million hectares. On many of these farms, the diversity and population of animal species is no less than that in official wildlife reserves. The use of wild animals on private land has been allowed in Namibia since 1967, and in the interim, the populations of large wild animal species have grown by about 70 percent.

Most international conservation organizations now regard sustainable use as a meaningful approach to species protection. Unfortunately, there is still some resistance to this approach in some "first-world" donor countries. This could prevent organizations providing aid to development and conservation organizations from fully supporting projects for the use of animals living in the wild. In the affected regions, however, the loss of natural habitat is driving wild animals into an ever-shrinking area, and we must discover and eliminate the reasons for the loss of habitat in order to stop this trend.

On the other hand, in many areas densities of wild animals are too high and can completely destroy vegetation and transform wilderness into wasteland. A prudent and scientifically supervised reduction of the number of animals in areas that are simply overpopulated not only protects habitats but also benefits all the other species living in them.

Projects for sustainable use have been particularly influenced by two international treaties: the Convention on International Trade in Endangered Species (CITES) and the Convention on Biological Diversity (CBD). CITES was adopted in 1973 out of concern for the fauna and flora whose numbers have rapidly decreased as a result of international trade. Although in the 1970s and 1980s CITES operated through prohibitions on trade, chiefly in accord with the principle of prevention, today efforts are being made to find a reasonable balance between prevention and sustainable use. In its first appendix, CITES lists all the endangered species in which trade is forbidden, and this ban holds for all the products made from these life-forms as well. However, a trade ban is an effective means of protection only when it is in fact international trade that is threatening a species.

The key to sustainability may be not in trade bans but rather in trade controls: We need a system that encourages trade in products of sustainable use and at the same time penalizes illegal trade in animals and plants that are poached. CITES seeks to achieve this by classifying species in its second appendix. Species appearing on this list may be traded only if the seller can prove that sustainability has been ensured: The export of an animal or plant must not endanger the survival of a species. For countries with management that works, and where this protective kind of use functions, CITES provides for corresponding special regulations.

The most important international treaty concerning natural resources is the Convention on Biological Diversity. It emphasizes the importance of sustainable use of genetic resources and a fair distribution of the proceeds that result from such use. The CBD offers an appropriate platform for the development of measures in both national and international conservation that view sustainable use as an incentive for protecting biodiversity.

However, modern conservation, which seeks to make sustainable use a viable alternative way of using land, still concentrates mainly on the microeconomic (that is, the regional and local) level. It disregards the fact that macroeconomic (that is, national and international) forces undermine this approach. Such hurdles have to be overcome before sustainable use can realize its full potential and achieve success. We have to do away with market distortions that promote – through inappropriate subsidies, taxes, and superfluous rules – destructive ways of using land.

Putting an end to competitive distortions, promoting legal marketing of natural resources, and creating markets for animal products are, however, not the only ways of making wild animals more valuable for leaseholders and landowners.

Positive political incentives should also be used. Specifically, only land use that directly serves the preservation of biodiversity should be promoted. The incentives include conservation subsidies for the maintenance of species living in the wild and their natural habitats. Examples are the subsidies used in the European Union's "set aside" program or differential land use and development taxes. Another possibility would be flexible tax rates for different forms of land use and development, depending on the degree to which they damage nature.

UNEP will certainly seek to avoid distortions in the market economy and in policies at the international and national levels, and support measures that promote sustainable use of genetic resources. We want to help developing countries create legal frameworks that will make it possible for village communities to benefit from the economic value of their natural resources and at the same time maintain the value of the ecosystem. Of course, UNEP must also ensure that the principles of protection and sustainable use are kept in balance – and always with a view to the welfare of the animal population.

Sources for text in margins:

1 R. Bailey: The True State of
the Planet, 1995
2 W. Krug, Geographische
Rundschau 5/1999

The opinions represented in this article are those of its author and not necessarily those of UNEP.

Sea bream; mackerel; trout

Wild Economics: Calculating Nature's Worth

If insects pollinate fruit trees, if beautiful landscapes attract tourists, and if forests purify the air, then biodiversity produces profits amounting to billions of dollars. Environmental economists calculate the value of plants and animals: How much does the world cost?

For seven and a half million years, Deep Thought reflected on the most ultimate questions: the meaning of life, the world, the universe, and all the rest. Wise programmers fed data into it. Unceasingly, the superbrain's circuits whirred and clicked. And for seven and a half million years, the inhabitants of Magrathea longed for eternal truth. Then came the big day. Deep Thought cleared its throat and announced: "Forty-two!" Finally, the "Great Answer" was given. Unfortunately, however, no one who knew precisely what the question had been was still alive.

Economists' efforts to calculate the monetary value of the planet Earth somewhat resemble the labors of Deep Thought, the computer in Douglas Adams's science fiction best-seller series, *The Hitchhiker's Guide to the Galaxy*. They appear simultaneously heroic and futile. Thirty-three trillion dollars – that was the answer Robert Costanza (University of Maryland) gave to the question, "How much does the world cost?" He added up the contributions made to humans by all ecosystems and came up with that sum: 33 followed by 12 zeroes. In dollars. And per year. Costanza made his calcula-

Wild animals and domestic animals (Sudan)

BANK OF SUDAN

FIVE SUDANESE POUNDS

tions with the best of intentions; he wanted to show the public the inestimable value of nature. But one figure puzzled the astonished audience: If we calculate all the services and goods produced by all 200 countries in the world, we end up with slightly more than half of the 33 trillion. The global economy currently produces no more than that. So what about the rest?

Is nature uncountable, then? Are the services that it provides not only free but also impossible to evaluate in dollars and cents? Without monetary equivalent, without cost – without value? On the contrary. Let us imagine that our planet has been transformed into a corporation: Blue Planet, Inc. Its range of products would include potable water and clean air, fish and meat, romantic sunsets at sea, and birdsongs in the morning. In addition, it provides storms, floods, and erosion, and uses trees and mangroves in connection with these. Its service team includes pollinators (insects), waste disposers (fungi), and ventilators (algae). And bacteria as jacks-of-all-trades. For investors, Blue Planet, Inc., is an attractive enterprise. It has much to offer. Everything, in fact. Everything that humans need and on which consumers are dependent for their existence. This drives up prices. How much are you prepared to pay for your next breath of air?

No economist would need to take the trouble to calculate the value of nature – were natural habitat not what is most often sacrificed to Mammon. Nature is converted into fields and plantations, settlements and industrial zones, because the new use promises to be more profitable than the original condition. Yet it is becoming increasingly clear that if the calculation does not include the natural economy, the long-term economic accounts will not be in balance. Damage will be greater than use.

It is time that analysts realistically assess Blue Planet, Inc. This is not a matter of ultimate questions and "Great Answers." More helpful would be practically oriented calculations indicating how to make the protective use of nature a successful strategy. The young discipline of environmental economics has made such a comprehensive view its own. Two American scientists ask how much insects are actually worth. The economic value per beetle – for exam-

Snowy owl
(Canada)

173

ple, the value of the chitin in the exoskeleton – is minimal, and the aesthetic value is a matter of taste: Insects are not exactly cute. But insects are unique in the realm of animals by their variety – 950,000 species have been scientifically described.

Insects, including beetles, develop along with the flowering plants with which they have established mutually beneficial partnerships. Many beetles feed on nectar, and in return they help pollinate 90 percent of all flowering plants. Without these winged gardeners, flowers would not reproduce. Orchid meadows would disappear, tropical forests would become barren, carpets of water lilies would wilt. But their contribution is not limited to pollinating wild ecosystems – most cultivated plants (with the exception of grains) are also dependent on animals for pollination.

Bees help potatoes, cassava, soybeans, peas, and sunflowers thrive, help fruit trees such as plums and peaches produce rich harvests, and make cabbage and garlic flourish. These animal farmhands are punctual, reliable, relatively undemanding, and do not take vacations. And they never go on strike – at least not for higher wages.

In fact they deserve a rich reward. The value of their pollination to agriculture can be calculated. The brothers Lawrence and Edward Southwick – one an economist, the other an agronomist, an ideal combination for environmental economics research – figured out how much money honeybees "earn" by pollinating sixty-two species of crops; as a standard of comparison, they used other means of pollination. For the United States alone, the brothers Southwick concluded that bees save consumers $6.7 billion every year, a sum that is reflected in lower food prices.

It might be objected that this amount is purely theoretical, a mere intellectual game in the style of Deep Thought. But it becomes clear, when bees are absent, that this is not so. Toward the end of the 1990s honeybees died in large numbers in the United States as a result of an epidemic of mites carrying a disease that infected the bees. Profits from the fields dropped, and the prices of some foods increased significantly. Insects are like car keys: You notice them only when you've lost them.

Wolves
(Lithuania)

Oscar Wilde once defined a cynic as "a man who knows the price of everything and the value of nothing." Today, the same suspicion hangs over economists who are trying to put a price tag on the previously unacknowledged services of Blue Planet, Inc. There is no doubt that plants and animals have an inherent value and therefore, from a philosophical point of view, have a primordial right to exist. But what should we do when politicians, planners, and managers dig in their heels? If some people understand only the language of credits and debits, then we have to present the ecological bill.

Take the case of New York City. The mayor used to be able to boast that "the champagne of drinking waters" flowed from Gotham's faucets. By the early 1990s, however, the quality of the water was rapidly declining. What happened? Traditionally, the Catskill mountains north and west of the city provided good water. Forests sucked up the rainwater like sponges and sent it on a cleansing voyage to New York. Nine million Gothamites drank and enjoyed this water – and the service provided by Blue Planet, Inc., was taken for granted. But clean water does not simply come out of the faucet. Over the years, an increasing number of dairy farms and small industries had been established in the Catskills, mainly along the rivers. Towns grew, forests shrank.

The day was not far off when New York was no longer going to meet the Environmental Protection Agency's strict guidelines for water quality. Then a filtering system would have to be installed, and it would cost $8 billion to build and another $300 million a year to run. Good old conservation offered a more cost-effective alternative. Scientists determined that it would cost only about $2 billion to clean up the watershed. Spread over ten years, this would represent a savings of $9 billion, and that was pretty persuasive. In 1997, the city began buying up land along the rivers and reservoirs, renovating old purification ponds, and more closely monitoring effluents from factories. New York made up its mind; instead of spending money, they were going to make greenbacks grow in the Catskills.

This kind of investment in the well-being of Blue Planet, Inc., could also be attractive to private investors. For instance, American economists Graciela Chichilnisky and Geoffrey Heal (Columbia

Springboks
(Namibia)

175

University) suggest that "a city like New York could sell shares in a purification project. Part of the money saved by choosing conservation over the construction of a purification system would be paid to shareholders as a yield on their investments." By issuing such "green stocks," communes could not only interest their citizens in conservation, but at the same time avoid debt. Scientists cannot often come up with a balance sheet as unambiguous as in the case of New York's water supply. Will a strip of coastal land be more profitable as a mangrove forest or as a shrimp farm? As mud flats or as an oil field? As sand dunes or as a marina? "There are no objective values," says Professor David Pearce, the director of London's Centre for Social and Economic Research on the Global Environment (CSERGE). "We are not trying to dictate what value humans should assign to nature. However, as scientists, we can use the appropriate instruments to find out what priority a society accords a specific good." Pearce points out that it was the growing awareness of the environment that produced what people in the field call the "willingness to pay." At the same time, the galloping destruction of natural habitats tipped the scale: Scarce goods are expensive goods.

Among eco-economists, David Pearce is considered an international expert. In the 1970s there were barely two dozen specialists in his discipline in Europe; now there are about 1,000. Their job is to analyze the competitive battle that is occurring in many places: a battle between those who are destroying wild areas and those who are protecting them. Explaining the boom in his discipline, Pearce says, "Nature is losing everywhere. Our expertise is in such great demand because we can say why. The answer is: Nature has no economic value for the people who live near it. It only costs them." He gives the example of an elephant herd in Kenya: If it tramples down a farmer's millet field, it threatens his harvest; if it approaches his hut, it threatens his life.

That is why Professor Tim Swanson, Pearce's colleague at the CSERGE, sees his most important job as developing mechanisms through which wild animals can provide the local population with cash. "Only in that way does the local population have an incentive to maintain wild areas over the long term," Swanson says. This ap-

Tiger
(Vietnam)

176

proach is known as "making nature valuable," and it is not content with assigning fictitious price tags: Protecting species has to generate money.

For instance, take the pandas that live on the Wolong reserve in China (Szechuan province), where Tim Swanson did research. "The reserve is a source of all sorts of annoyances for the local people. They can no longer hunt there, or cut timber, or plough up fields. None of them loves pandas just because they're so cuddly." Guns are no more capable of protecting the reserve than are campaigns to inoculate the farmers with "ecological consciousness." Bertolt Brecht's remark to the effect that eating comes before morality is confirmed every day in Szechuan and elsewhere. How, then, can use be combined with protection?

For Swanson, the answer is quite easy in the case of the panda: "In fact, the animals are walking bank accounts. There are enough tourists who are willing to pay more for a visit to the poster animal of species preservation." Using opinion surveys, Swanson calculated the precise amount of this "willingness to pay," and recommended as a first step that the park's entrance fee be raised from seven to twenty-five dollars for foreign visitors. Combined with a well-organized panda tourism program, the reserve could take in $40 million a year. For Swanson, the second element of his strategy is still more important: "Part of the income has to go to the surrounding villages. When their residents learn that pandas provide advantages for them, poaching will decline as well." Swanson believes that the love of animals passes through the pocketbook.

For David Pearce, the earlier approach to conservation through prohibition has failed: "What people would like to do is to follow the motto 'Fence and forget': fence off the areas, drive people out of them, post armed guards at the barriers, and then everything is fine. Except: it doesn't work." Many of the protected areas, especially in poorer countries, exist only in name: paper parks. They are coming under increasing pressure. As the population around them grows, the hunger for meat, firewood, and new land also grows.

Most wild animals lives outside reserves, anyway, and thus in guaranteed uncertainty. They move on land that might at any time

Cranes
(Japan)

be transformed and lost as habitat. For them, conservationists coined the slogan "Use it or lose it." Only when raising wild animals is more profitable than agriculture will biodiversity be preserved. Over recent decades, this modern form of use and protection has achieved surprising successes. The CAMPFIRE project, begun in Zimbabwe in 1986, has earned worldwide acclaim. Its simple but ingenious idea was to privatize the country's wild animals, which had previously been under government control. The districts where buffalo, elephants, zebras, antelopes, lions, and crocodiles were indigenous were granted property rights. They were allowed to use a small number of the animals. The villages decided whether to eat the animals themselves or sell them to traders, whether they preferred to set up photo safaris or sell licenses to hunt big game. The quota was determined by ecologists and supervised by rangers. It varies from 1 percent of the population in the case of elephants to 6 percent in that of zebras. Only a fraction of the natural population growth is thus skimmed off.

Seen as private buffalo or community giraffes, the attitude toward African fauna has strikingly changed. Elephants are beautiful creatures but poor neighbors. Farmers liked them no better than rats in their larders. Africans found it increasingly difficult to understand why they should pamper them just because rich people in the Northern Hemisphere had taken a liking to wild animals. With CAMPFIRE, however, coexistence is worth the trouble. The more the animal populations grow, the greater the surplus that goes to the villages. They earn the most from trophy hunting. For an organized leopard hunt with beaters, tourists plunk down as much as $15,000. Protecting wild animals has become more profitable than raising livestock; or, as Chief Kanyurira from Matoka puts it, "Buffalo are now our cows." With the profits, CAMPFIRE villages are building schools, water supply systems, and clinics. And hiring rangers, who keep a watchful eye on their four-legged pots of gold. Many former poachers have become rangers.

CAMPFIRE is as highly respected by environmental economists as by the conservationists of well-known organizations such as the WWF or the IUCN. However, it is strongly opposed by so-called

Lizard and butterfly (Sri Lanka, formerly Ceylon)

animal lovers, most of whom live in industrialized countries. They cannot see big-game hunters as anything but monsters. Simon Metcalfe, the project's spiritual father, replies: "If you want there to be any rhinoceroses tomorrow, privatize them!" Well-controlled trophy hunting should not be taboo. Ultimately, we want to preserve not individual animals, but entire species. The dead elephant finances the future of its living relatives. Moreover, the elephant population, which is rapidly growing in southern Africa, has to be regulated anyway; otherwise it would destroy its habitat and drive out other species, as well.

The bleeding-heart opponents of all hunting are not saving the elephants – not even if their feelings sometimes lead them to make cash contributions. Here we see the difference between sentimentality and true love. "It is not enough," Tim Swanson says, "to sponsor an elephant for a year. It will live for sixty years. And the farmers have to deal with it all their lives. Only when they can be sure that the flow of money that they receive from wild animals will continue, will they start protecting them." No one would expect that paying for one night in a hotel room for someone would ensure that the room would be kept free for him for a whole year.

If gnus and zebras become sources of money, the analysis of the costs and uses of wild animals becomes more favorable. A comparative investigation of 140 farms in Zimbabwe showed that cattle raising produced an average profit of 2.5 percent, whereas specializing in wild animals produced 8 percent. This kind of yield encourages other countries to think about natural resources:

– More and more South African farmers are selling their domestic livestock, taking down fences, planting trees, and specializing in game farming: rhinos instead of cattle. The phototourism business is lucrative. Moreover, the rhinoceros populations have increased so much that South Africa can export these giant animals, though a bull will cost you $25,000.
– In Namibia, which also privatized its wild animals, the populations grew by 70 percent in two decades. A third of Africa's cheetahs romp on private farmland.
– In the United States the alligator population decreased radically

Gorillas (Democratic Republic of Congo, formerly Zaïre)

during the 1950s and 1960s. Thanks to a conception of protection and use that includes trade in skins and hunting licenses, alligators are again living all over their original range in secure populations.

– In the Royal Chitwan National Park (Nepal), rangers found it increasingly difficult to end poaching. While the human population along the park boundaries tripled, the rhinoceros population dwindled. The situation improved when villagers started receiving part of the park's income. Conservationists also helped villagers get into the souvenir business.

Is it permissible to kill butterflies? In Papua New Guinea, they are protected by a clause in the country's constitution – and they represent an important source of income. Worldwide, trade in lepidoptera amounts to $100 million a year, and Papua New Guinea is one of the biggest exporters of particularly large and colorful specimens. Illegal collecting, however, had brought many species – such as the largest, the Queen Alexandra's birdwing – to the brink of extinction. Then steps were taken to better organize and supervise trade. Collectors were encouraged to establish lepidoptera gardens, in which plants particularly rich in nectar (for the moths and butterflies) and food plants (for the caterpillars) were grown. Trees provided shade and protection from predators. Many previously rare and endangered species once more proliferated. However, the most important step forward occurred in people's heads. At one time the rain forests had been seen as good only for clearing and burning, but now it pays to maintain them; they have become a place where money grows on trees.

Conservationists achieve similar results, even when they not only tolerate long-term use of forest resources such as plants, fruits, and rubber, but even promote it. This is the best way to oversee and prevent any damage to the ecosystem. Without such monitoring, the concept of "protection through use" would be much too risky. Even in the case of ecotourism – an especially protective way of "harvesting" natural resources – care must always be taken to determine how many visitors an area can bear without doing long-term harm to the animals.

Hornbill
(Indonesia)

Blue Planet, Inc., has something to offer in the tourism sector. For the first time in history, more people live in cities than in rural areas. And these city dwellers want to go, as far and as often as they can, back to the land. Tourism has become the largest sector of the world economy, with 600 million foreign and 2 billion domestic travelers a year. And within this booming sector, travel to natural attractions is the area that is growing fastest.

It is hardly surprising that countries in the Southern Hemisphere, which are rich in species and poor in cash, have set out to attract this kind of tourism. In at least five poor countries (Kenya, Ecuador, Costa Rica, Madagascar and Nepal), it is the most important source of revenue. Conservation benefits from this, since without paying guests it would be impossible to finance many of the 12,413 large wildlife reserves around the world. Tourists transform wilderness into prosperity. Environmental economists have determined that one hectare of land in Kenya's Amboseli National Park brings in forty dollars a year, whereas a similar area used for agriculture brings in only one dollar. Studies show that a single lion brings in $27,000 a year – not a bad wage.

Marine biologist Charles Anderson has compared two hotly competing ways of using natural resources in the Maldive Islands: sharks for the frying pan and sharks as subjects for photographers. The market value of a gray reef shark is about thirty-two dollars. Scuba divers who travel long distances to observe and photograph the reef predator in its natural surroundings pay several times as much. The diving fee is thirty-five dollars, and this adds up to more than $17 million a year – a valuable asset for this poor island country. When the gray reef shark started disappearing at Fish Head – one of the most important observation sites – the divers started staying away, too. These twenty dead sharks caused, according to Anderson's estimates, a decrease in income amounting to $500,000 a year.

Wolf Krug, a German environmental economist who also works at CSERGE, sees the habitat of the mountain gorillas in the Virunga Mountains (in Rwanda, Uganda, and Congo) as another "gold mine." Together with the Dian Fossey Gorilla Fund, he is urging the governments concerned to improve the organization of animal

Lions
(Kenya)

tourism, which, his experience suggests, could bring in $60 million a year (far more than it now does). "The local population must absolutely benefit more from this financial boon, so that the gorilla area is not gradually destroyed," Krug says. Mountain gorillas in Rwanda, giant sea turtles off the Galapagos Islands, cranes in the Extremadura region of Spain, whales off Baja, California – as tourist favorites, these natural attractions are now right up there with the Mona Lisa, the Coliseum, and the Great Pyramid. Naturally, there is a danger that they – like the cultural magnets – may be overrun. But conservation has learned from hard experience, and has developed a whole arsenal of methods for controlling visitors. The size of visitors' groups is limited, paths lead around delicate zones, and people are taught how to behave.

Increasingly often, environmental economists are involved in management. Their task is to determine the optimal entrance fee, high enough to cover the reserve's costs, but not so high as to keep visitors away. Just a few decades ago, hardly anyone would have imagined that so many people would undertake long and expensive trips merely to see wild animals. But since wealthy countries have reduced their own large fauna to tiny residual populations, the desire has grown to see animals not only in zoos but also in the wild. And even those who stay at home are willing to make a small contribution. They just want these animals to exist. Environmental economists call this "existence value."

Funding for conservation is increasing in industrialized countries, and David Pearce sees this as evidence of a growing awareness of the value of natural resources. In 1990 the thirty largest United States conservation organizations alone took in about $273 million. Yet studies at the CSERGE think tank in London have shown that the true potential is much higher. "Citizens have realized," Pearce says, "that the maintenance of biodiversity is a matter of interest to the whole world. Thus the whole world must pay for it."

Over recent decades new mechanisms have been invented for transferring money from the Northern to the Southern Hemisphere in order to preserve biodiversity. One possibility is the so-called debt-for-nature swap: Conservation groups buy up at discount

Reed warbler
(Cape Verde)

prices the debt of owing countries as well as large amounts of biodiversity. In return for debt relief, governments promise to protect certain areas and provide wildlife parks with adequate budgets. Since 1987, about $120 million in debt have been exchanged in this way. The first swap was a deal between the American organization, Conservation International, and the government of Bolivia; in exchange for debt relief, the latter undertook to set up the Beni Biosphere Reservation. Protection grew, debt shrank.

In 1990 the Global Environment Facility (GEF) was established. This fund was also based on the conviction that developing countries should not be the only ones to pay for something that benefits the whole world. Consequently, the main contributors are the rich countries of the Northern Hemisphere. So far, $730 million have come in, almost half of which has gone to fund projects for preserving biodiversity.

So-called joint implementations are another way of distributing responsibility more fairly. The basic idea, formulated in the International Climate Convention's Kyoto Protocol, is that countries should finance environmental improvements where they get the greatest results for their investment. So in Scandinavia, air quality can be much more effectively improved by cleaning up outmoded industries in the Baltic countries than by filtering out the last bits of dust at home. The same goes for carbon dioxide; countries can also meet the Climate Agreement's goals for reductions by paying for large-scale reforestation in tropical areas.

David Pearce calls this kind of exchange – which also benefits biodiversity because it increases forest area – "exotic carbon." "In this deal there are only winners," he says. He welcomes any idea that creates a market for the fictive Blue Planet, Inc. "Anyone who wants to give money for conservation must find easy ways to pay it." He is very enthusiastic about a U.S. airline's idea of introducing "green tickets." They are to show two prices: a mandatory price and an optional supplement, the carbon dioxide component. For this additional payment, trees will be planted that absorb precisely the amount of carbon dioxide that the corresponding flight produces. The passengers fly away in good conscience, knowing that

Sea turtle
(Aruba)

they are leaving nothing behind them in the skies other than two white contrails. However, this plan has not yet gone into effect.

As is shown by the example of forests as a CO_2 sink, Blue Planet, Inc., provides many of its services in a hidden way, as a secret helper of humankind. For a long time it was not recognized, but as awareness of its value increases, more attention is being paid to it. Nature is not only a very efficient producer but has also developed a broad range of technical ideas that – in contrast to many human pipe dreams – have proven their effectiveness. And they did so long ago; modern engineers looking for new solutions are copying nature's age-old inventions.

Leonardo da Vinci took the bird's wing as a model when he designed mechanical sailing ships and gliders. Today, scientists are systematically investigating natural functional principles in order to use them commercially. Materials experts, chemists, architects, and automakers are rummaging around in nature's toolbox. This new direction that biology and technology are exploring is known as "bionics."

For instance, aeronautical engineers are suddenly taking an interest in sharks. These predators swim with amazing speed and grace. They do not owe their agility solely to a particularly smooth skin, as landlubbers are prone to think. On the contrary, a shark's whole body is covered with tiny hooks that make it as rough as sandpaper and thus decrease the resistance of the water (the asperities prevent the formation of vortices that can slow forward movement). Engineers imitated this idea by producing a foil with similar qualities. When they put this artificial sharkskin on passenger jets, they were able to decrease fuel consumption by 30 percent.

German botanist Wilhelm Barthlott observed that water pearled on lotus leaves as if they had been oiled. No dirt stuck to them. The reason for this self-cleaning ability was the microstructure of the leaf surface. Thousands of tiny wax-crystal bumps prevented dirt from adhering. Barthlott applied for a patent on a procedure for producing an artificial material that produced the "lotus effect." Cities plagued by graffiti showed great interest in Barthlott's invention because graffiti could easily be removed from the walls of buildings that had been treated with the material.

Zebras
(Rwanda)

Developing new biomolecules is a difficult business. For instance, it takes years of labor to produce polymers that combine the advantages of wool with those of artificial fibers. But not if one understands how evolution arrives at new models – through mutation and selection. Biotechnology professor Peter Schuster makes these principles his allies. In his laboratory in Jena, a town in Germany, he has automated machines constantly hatching new molecular variants, looking for the ones that happen to have the desired abilities. He has made evolution race ahead in test tubes.

These are only three of hundreds of examples of how technologists take their inspiration from nature's creativity. Materials researchers put spider webs under the microscope in order to discover the secret of their tensile strength. Scientists at the Massachusetts Institute of Technology use insects as prototypes for crawling robots and "intelligent prostheses" that can make life easier for the handicapped. Statisticians learn from the skeletal structure of grass stems. Packing material engineers are looking for materials that preserve their contents, are durable, and leave no garbage behind them after use; among birds, animals with chitinous exoskeletons, and banana skins, that has long been the technical standard.

Another way of using the richness of nature's insights is to put living organisms directly to work. In this way, technonological conservation is achieving amazing successes:

- In North Carolina, purification systems were developed based on bamboo plants. In bamboo's root systems live organisms that filter harmful substances out of the water and transform them into food for the quickly growing bamboo itself.
- Phytoremediation is a term for a process using plants to purify contaminated soil. For instance, Indian mustard (*Brassica juncea*) has the property of absorbing lead and accumulating it in its cells. During its growth period, it actually sucks heavy metal out of the earth. It is much less expensive later on to deal with the resulting biomass as a form of hazardous waste than to excavate whole areas. Similarly, sunflowers have been used to help clean up the radioactively contaminated region around Chernobyl.
- Bacteria are intentionally put into coastal waters in order to clean

Shark
(Cook Islands)

185

up the sea after a spill from a tanker carrying petroleum. They eat their way through the oil slick and excrete harmless substances. Other microbes are even capable of breaking down dangerous compounds such as DDT.

– German surgeons have put *Lucilia sericata* to work: These large maggots are attached to wounds in order to remove dead tissue. Especially in the case of diabetics, whose wounds do not heal well, the physicians have used this method to achieve striking results.

– Bees can help locate mines; this bold idea is being pursued by American chemists working with insect experts from the University of Montana. Landmines discharge small amounts of the explosive TNT into their surroundings. Bees go out to collect nectar and pollen, and when they come back, they can be chemically analyzed – so that honey collectors become heralds of peace.

As a puzzle solver, nature has a decisive advantage over humans: time. It has been carrying on its experiments for 3 billion years. In doing so it has created a complexity of genes, molecules, and biotopes that the human mind still is not able completely to understand, let alone reproduce. This diversity and complexity are also the reason that managers and politicians are showing more interest in ecological processes. Increasingly, they find themselves trying to operate in systems – whether markets or governments – that are becoming constantly more complex – closely interconnected, affecting each other in chaotic ways, and never in balance. Just like natural communities. Hence a sort of "business bionics" is developing as an attempt to copy the success of Blue Planet, Inc.

"Leading figures who are learning from ecology and evolution," writes James Moore, an American business commentator, "are arming themselves with wise strategies for taking control of the future." According to economist Brian Arthur, markets function very much like ecosystems. He argues that companies should no longer be seen as well-lubricated machines, but rather as living organisms that are controllable only to a limited extent – so that many managers' feelings of omnipotence are illusory. Another lesson one can draw from this is that the survival of the fittest does not mean tearing one's competitors limb from limb. It is much smarter to seek

Products made from plants and animals are used in many different ways. 1

Market Prices of Various Goods Derived from Plants and Animals	
Consumer price per kilogram or liter, in U.S. dollars	
Human growth hormone	20,000,000
Vincristine (used in treating cancer)	11,900,000
Cocaine	150,000
Lear's macaw (alive)	24,000
Dried bear's gall	7,000
Saffron	6,500
Tiger bones	3,000
White truffles	650
Cotton	1.5

partners and to build networks that offer advantages to everyone involved. Biologists call this "symbiosis."

Jared Diamond, an American evolutionary biologist, draws still another lesson for human economic activity. His key thesis is that differently constituted "biopackets," the natural resources with which each major region is endowed, have had a decisive influence on the course of human history. In Eurasia, the biopacket was most favorable to the development of agriculture, and that is why high cultures were able to emerge there at an early date. The continent got a powerful head start.

The situation was quite different in America and in Australia. When white colonists arrived, they found indigenous peoples who, for the most part, were still living in the Stone Age. How did this temporal difference come about? The answer once again is connected with the biopacket. When the first humans settled North America and Australia, the large mammals there died out in a relatively short time, either by mere coincidence, or as a result of hunting. In any case, with them disappeared the chance of moving into agriculture and the raising of livestock: There was simply a lack of the raw material for domestication and breeding. Incapable of successfully resisting the white conquerors, the indigenous peoples were almost wiped out.

This tragic historical development represents a textbook case of humankind's treatment of nature. Cultures obviously flourish most when they use their biopackets without plundering them. Becoming a shareholder in Blue Planet, Inc., pays long-term dividends. Edward O. Wilson, a sociobiologist and a famed researcher in the field of biodiversity, also argues for a wise economic use of biodiversity: "Undisturbed nature is like a magic spring: The more we make of the reservoir of knowledge and possibilities for use that it provides, the larger it becomes."

See also plates:

Forests as an Economic Factor
Page 150
One Tree, Many Products
Page 152
The Green Pharmacy
Page 156
Nature as a Productive Force
Page 158

Sources for text in margins:

1 K. ten Kate, S. A. Laird:
  The Commercial Use of
  Biodiversity, London 1999

Fair Dealing: Who Gets Nature's Dividends

Plants, animals, and microbes provide valuable ingredients for the pharmaceutical industry. Source countries and users are struggling to determine fair rules governing patent rights and the transfer of royalties.

For ten hours, Fritz Kamti has been digging holes in the sandy soil of the Kalahari Desert. Day after day, from March until June. Sweaty work, in heat reaching 105° Fahrenheit in the shade. There's no shade to be found anywhere in this part of Namibia. The sun beats down without mercy on the barren thornbush savanna in which Fritz and other members of his clan are working. Equipped with machetes, spades, and spoons, they are digging up the roots of the grapple plant (*Harpagophytum procumbens*). To judge by what one can see of the plant above ground, it doesn't look like it would be worth the effort – withered, inconspicuous branches with a few little, dried out leaves just above the ground, hardly different from all the other plants in the area; here and there claw-shaped fruits hang from the plants.

Fritz owes his first name to Namibia's German colonial past. He is a San (a "Bushman," as the Europeans somewhat disrespectfully said) and a village elder in Vergenoeg, a hamlet whose name does not appear on any map. His grandfather showed him that it was worthwhile to dig up the thick roots of the grapple plant, which run as deep as a meter below the surface. Shaped like cucumbers, they are cut into thin slices and spread out to dry in the sun for two days. Dissolved in hot water to make a kind of tea, for ages they have been helping the San relieve stomach and limb pains. Yet for a few years, Fritz and his companions have no longer been working to supply their own needs. Dealers from Namibia's capital, Windhoek, and the neighboring countries of Botswana and South Africa, as well as scientists from Germany and plant experts from Great Britain, come to Vergenoeg. The grapple plant is doing well. The San's concoction has become a valuable raw material for pharmaceutical research.

Even before World War I, a German soldier serving the colonial government in Namibia had heard about the healing properties of the grapple plant and repeatedly observed San medicine men using it. Later on, he was the first to sell the roots as a "wonder cure and pharmaceutical treasure" in Germany.

A series of clinical studies have recently confirmed the effectiveness of grapple plant roots; no negative side effects have been reported. In the interim, trade has sharply increased; in 1998, Namibia exported 500 tons of dried grapple roots, a significant increase. They are sent primarily to the European Union, where an extract from the roots is sold in the form of pills used to fight rheumatism. Preparations made from grapple plant roots have become a big seller in German pharmacies and drug stores. The United States might soon follow suit, since research there has recently suggested that the roots might be able to decrease cholesterol.

Jules and his extended family from Abrotsi, Côte d'Ivoire, walk for several hours until they reach a river that can be crossed only in a narrow dugout canoe. After paddling for an hour, they finally reach the other side. In the village's own rain forest they look for the bright red fruits of a palmlike plant that does not attract the attention of those who do not know it, because it grows everywhere in the area. The natives call it *katemfe* or *ndebion*, while European explorers named it *Thaumatococcus daniellii*. For 10,000 people in Côte d'Ivoire, this jungle plant is a source of income. Its fruits contain the sweetest material in the world; it is 2,500 times sweeter than sugar – and it has virtually no calories.

Jules's ancestors used the fruits to sweeten their food and beverages. This traditional use had long been forgotten. But in the 1970s, European food chemists, looking for natural sugar substitutes, rediscovered *ndebion* fruit. Since then, representatives of large companies have been touring West African villages and buying the fruits natives have collected in the jungles. There is a positive side effect: Since people have been making money from forest fruits, there has been a decrease in burning forests and cutting them down to clear land for agriculture.

In a number of processing centers in Côte d'Ivoire, the fruits are opened by skilled women who remove the seed-bearing, white upper part, which is then deep-frozen and shipped to England. All one has to do is lick the seed, and hours later, lemons still taste as sweet as sugar, and grapefruits lose their acidity. In Europe, the fruits are dissolved in acidulated water and the supersweet protein thaumatine is filtered out. It is used in more than 100 products – from fruit juice to potato chips and coffee products – chiefly on the Japanese and North American markets. Only in tiny doses, of course, since one kilo of thaumatine costs 5,500 U.S. dollars and has as much sweetening power as two and a half tons of sugar.

Catharanthu roseus
(Madagascar evergreen)

Debaprasad Chattopadhyay is one of the microbiologists rummaging around in rain forests, looking for animal and plant species previously unknown to the scientific community. His sponsor is an Indian government institute. Always hoping to find something that will help fight cancer, AIDS, or other scourges, he makes use of indigenous peoples' knowledge of medicinal plants. He questions medicine men, peers into shamans' kettles, listens to stories told by village elders.

In 1993, during an expedition in the thick rain forests of the Andaman Islands off the east coast of India, he encountered the Onge tribe and was amazed to find that although they were positively surrounded by mosquitoes, none of them had fallen ill with malaria. After he had won their trust, he asked why there was no malaria among them. They took him to a small, smoke-filled hut where there was a pot full of a bitter medicinal drink. The Onge gave Chattopadhyay several specimens of the plant that was used in making this preparation.

Back in his laboratory in Port Blair, on the Indian coast, Chattopadhyay made a sensational discovery: Two of the plants included fever-reducing ingredients, and a third reduced the number of malaria-causing agents in infected human blood. Soon thereafter he had an opportunity to test the effectiveness of the Onge's medicine on his own body. After making further visits to the rain forests, he himself became ill with malaria. He drank the liquid extracted from the plants and was well again in only three days. The episodes of malaria did not return. Convinced of the plant's curative properties, he got a local physician to administer the plant extract to seven patients who had the worst, most lethal form of malaria – and once again it worked.

These results are so intriguing that more precise studies should be done to test the effectiveness of the Onge's medicine, which represents a new chance to save the lives of the 2.7 million people who die of malaria every year. And it is also a drug that could be worth hundreds of millions of dollars.

Rauwolfia serpentina (Indian Snakerost)

Felipe also hopes to find a best-seller in the tropics. He is part of a team of some forty women and men who are working on behalf of the Instituto Nacional de Biodiversidad (INBio) in Costa Rica's national parks. He used to be a simple farmworker, but since he knows the Guanacaste area in northwestern Costa Rica like the back of his hand he was trained as a "parataxonomist," a species expert without an academic degree. His task is to study biodiversity and collect specimens in the region, the last tropical dry forest in Central America. This inventory of nature is supposed to provide Costa Rica's national park officials with an overview of the range of species that remain to be discovered and studied. The insect realm seems particularly worthy of attention, even though only a tiny part of its species is known to science.

Every day, Felipe spends several hours in the forest collecting caterpillars. Their strange appearance responds to ecological needs – bright colors, long barbs, or thick hair conspicuously warn predators that they

Genes and the fair distribution of profits
At the global summit in Rio de Janeiro in 1992, the Convention on Biological Diversity (CBD) was approved. The more than 170 signatory states were also concerned to ensure fair rules for the use of genetic resources and the resulting profits. Below is an extract from the convention:

Article 1
The goals of this agreement ... are the preservation of biological diversity, sustainable use of its component parts, and a fair and balanced use of the advantages resulting from the use of genetic resources.

contain poisons and that it would be better not to eat them. It is just these toxins that researchers see as very promising candidates for new drugs.

Felipe has already collected almost a thousand different species of insects. In a small shed not far from the Guanacaste National Park offices, he puts the caterpillars in plastic bags full of leaves. They are to grow thick and fat in the bags before he sticks them in the freezer. Information about where and when the caterpillars were found is put on every bag, so that more caterpillars can be found there at any time. Felipe types into his laptop information about any unusual modes of behavior that might be helpful to the researchers.

Hypericum perforatum (true St. John's wort)

Extracts from the caterpillars and stored data are sent to one of the world's largest pharmaceutical firms, Merck, in the United States. In the early 1990s Merck became the first firm to sign a contract with INBio, and it regularly receives thousands of samples of insect and plant materials from Costa Rica. In return, INBio receives a million dollars every two years. But not only that: If Merck succeeds in using the wealth of species in the tropics to make a new medicine against diabetes, viral diseases, or cancer, for instance, IN-Bio also receives up to 5 percent of the proceeds. The royalties could eventually amount, according to estimates made by the respected World Resources Institute in Washington, D.C., to more than Costa Rica's current export revenues for coffee or bananas.

Fritz, Jules, Chattopadhyay, and Felipe are part of a new, world-wide generation of hunters and gatherers: bioprospectors. These prospectors are looking not for precious metal, but for green gold; the known and unknown active agents in plants and animals are their capital. Traditional suppliers of plant-based drugs as well as the pharmaceutical, food supply, and cosmetic sectors profit from this capital. Major buyers also include modern companies in the life-sciences sector, which supply, in addition to medicines, seeds, pesticides, and diagnostic products. Without bioprospectors' knowledge and work, the use of natural resources on a grand scale is not conceivable.

For ages, people have been using nature's healing powers to relieve pain and to cure diseases. Today, natural substances are being used medically in two different ways. One of these is natural medicine, in which natural materials are used without chemical alteration. Most people in poor, developing countries have to rely on such means.

However, the use of medicinal herbs is also increasing in industrialized countries, not only in herbal medicine, but also in the production of cosmetics and food additives. "Bio" is in, since more and more consumers distrust products that come from test tubes. In Europe alone, 2,000 plants are sold for medical purposes. Germany is one of the world's largest importers of such plants, with a total of more than 45,000 tons a year. In 1994, $12 billion worth of medicinal herbs were sold worldwide. This market is growing faster and faster.

In the second way of using natural substances medically, researchers working for international chemical and pharmaceutical firms take natural materials as models for the production of new products. Computer-controlled technologies make it possible to investigate in a single day thousands of different samples for pharmacologically interesting materials. After their structure and modes of action have been determined, the substances are reproduced, either synthetically or in large-scale cell cultures, with the help of modern biotechnology and gene technology.

Gentiana lutea
(yellow gentian)

In every leaf, every caterpillar, every tree we find the results of millions of years of evolution – a cornucopia of unknown molecules and genetic material that can serve as a source of new active agents. The substances caterpillars produce to ward off their enemies, or the pheromones used by a male beetle to attract females – these are all genetic resources that could provide the starting point for a major business enterprise.

As Dr. Bennett M. Shapiro of the American firm Merck notes, "Among natural materials we encounter revolutionary substances that we have never seen before, that no chemist could ever imagine. Looking into what nature has produced in the way of chemicals gives us a glimpse of the universe of infinitely numerous chemical structures."

Article 2
In this agreement, "biological diversity" means the variability among living organisms of all origins, including, among others, terrestrial and marine or other aquatic ecosystems...; "genetic resources" means genetic material of actual or potential value.

We often don't look at the packaging, but even drugs that have become synonymous with the achievements of chemistry, such as aspirin, go back to natural active ingredients. Twenty-five percent of all prescription drugs sold in the United States contain major ingredients that were originally found in plants. In Germany, 50 percent of prescription drugs contain such ingredients.

As mentioned earlier, reserpine, a substance used to fight high blood pressure, originally came from the root of the plant rauwolfia. Taxol, a substance that has been successfully used to treat ovarian and breast cancers, was first isolated in the bark of the Pacific yew tree before chemists were able to reproduce it synthetically in the laboratory. Vinblastine and vincristine are among the drugs most frequently used to fight leukemia in children. They were derived from the tropical Madagascar evergreen, and the American firm Eli Lilly sells more than $100 million worth of these products annually. Experts estimate that thanks to increasingly refined techniques and more heavily funded search expeditions, trade in drugs based on plants could soon reach more than $500 billion a year. Although skeptical observers consider this figure exaggerated, one thing is certain: The nature business is booming as never before.

Thymus vulgaris
(true thyme)

Yet despite all the work of bioprospectors like Fritz, Jules, Chattopadhyay, and Felipe, the countries where very promising plant and animal material is being discovered are the ones profiting the least from it. Above all, developing countries need money and technology to make economic progress and to finance conservation.

Exploitation of natural substances can also occur at the expense of the environment. For example, in Namibia: Since the grapple plant's properties have been discussed among rheumatism-plagued Europeans, demand for its roots has increased enormously. Yet, for years trade in the roots was completely unorganized and uncontrolled. Fritz Kamti received the equivalent of less than fifty cents for a whole day's harvest. Buyers did not come regularly, and no collector knew when he could sell how much. The result was that in a short time the plants were completely overharvested. Fritz knew how to cut the roots so that they would grow back, but no buyer paid him more just because his harvesting methods were environment-friendly. The miserable conditions under which the harvesting took place has led to the grapple plant being wiped out in many parts of Namibia.

This is no isolated case; since trade in medicinal herbs has increased worldwide, these plants are in a critical situation everywhere. Customers who are not environmentally informed often have no idea that the expensive herbal cream from the beauty shop, the chemical-free tea from the health food store, or the biopills from the pharmacy are often anything but healthy for nature. About 90 percent of the species of medicinal herbs sold around the world do not come from regular cultivation, but rather from uncontrolled collecting in the wild. Whether it is arnica, yellow gentian, thyme, oregano, or Greek mountain tea, many of the best-known end products were ripped out, dug up, or cut down on a large scale. In Europe alone, especially in eastern and southern Europe, at least 150 plant species that have curative, cosmetic, or aromatic effects are endangered.

Chamomilla recutita (chamomile)

It is true that a few of the best-sellers, such as peppermint and chamomile, are produced entirely by cultivation. Others, such as St. John's wort, whose ability to relieve light depression has caused a sensation, are now being produced at least partly by cultivation. "Still, most medicinal herbs come from collecting in the wild," says Dr. Dagmar Lange, a biologist who keeps an eye on trade in medicinal herbs for WWF and for the German Federal Conservation Office. "Either they cannot be cultivated, allegedly, or actually contain more effective substances in their wild forms, or are simply cheaper to collect in the wild." Most of them come from China, India, Egypt, Chile, Pakistan, Mexico, Morocco, Bulgaria, and Poland.

Low rates of pay for collectors in eastern Europe and in the third world lead them to make unrestrained use of wild areas. Yet breeding and cultivation are not the only solution to the overuse of plants that grow in the wild. Not every kind of collecting is ecologically

Article 8

Each signatory to the accord will ...

g) Initiate or maintain means for regulating, administering, and supervising the risks that are connected with the use and release of living organisms that have been modified by biotechnology and that may have negative effects on the environment ...

j) Respect, protect, and preserve, within the framework of their intragovernmental legal regulations, knowledge, innovations, and customs of indigenous and local communities that have traditional ways of life that are important for the maintenance and long-term use of biological diversity ...

reprehensible. Correctly cut, only partly decorticated, or dug up in such a way that the main roots remain in the earth, plants can regenerate themselves. Dr. Uwe Schippmann of the German Federal Conservation Office argues for controlled collecting in the wild: "The revenues connected with this can become an economic incentive to maintain the environment of medicinal plants."

That is precisely what informed dealers in medicinal plants and concerned pharmaceutical companies are doing in Namibia. In order to be able to produce enough grapple plants in the future, a few firms are now buying only from collectors who use environment-friendly methods, and are prepared to pay more for them. With the help of biologists, efforts are being made to determine how many plants there are in Fritz's homeland near the border with Botswana, and how many can be harvested without threatening the population of this species. The maximum number of roots each collector can dig up is set in advance. Protective harvesting ensures that the plants can grow back within one year. However, above all, care is taken that Fritz and others in his village are better paid than they were before – they now get paid more than ten times as much – and can rely on the buyers' coming on a regular schedule. No one in Vergenoeg any longer has an interest in exploiting nature shortsightedly.

Nonetheless, this pleasant story about plants used to benefit patients, natives, and nature, all at the same time, is unfortunately an exception. The prospects for Jules and the others from Abrotsi in Côte d'Ivoire are less rosy. Gene researchers have been able to isolate the genetic material that produces the supersweetener in the West African plant *ndebion*, and to transplant it into other organisms. Soon, genetically altered tomatoes and head lettuce will produce the sweetener thaumatine in their fruits and leaves – not in Côte d'Ivoire, but rather in modern greenhouses in the United States and Europe. Scientists have already patented the corresponding know-how.

Through patenting, what is freely available in nature becomes private property. Molecular biologists working for the University of California and a small genetic technology firm are becoming the

Mentha piperita
(peppermint)

owners of all the genetically altered plants that contain the sweetener gene. If the sweetener made in the laboratory turns out to be cheaper, then the red fruits from West Africa will become superfluous. The plants produced in the North will replace those in the South, and the patent holders will cash in at the expense of the people in West Africa.

Even Debaprasad Chattopadhyay's discovery of a new way of fighting malaria did not have a happy end. He had dreamed of wealth and fame, which he planned to share with the Onge tribe. Fewer than 100 members of the tribe have survived in the wild, and their habitat is steadily becoming smaller. In order to provide them with a secure income for the future, Chattopadhyay wanted to share the income from his wonder drug with the natives – for he never would never have discovered the antimalaria plant without their botanical knowledge. But when he returned to the governmental institute where he worked, he found that one of his superiors was already planning to apply for a patent on the malaria drug – without, of course, giving any thought to sharing the proceeds with the Onge. When he was asked to name the plant for the patent application, Chattopadhyay declined to reveal the secret – in order, he said, to prevent the exploitation of the tribe and personal enrichment of his superiors at the institute. Since then, officials have forbidden him to pursue further research among the Onge or to carry out more pharmaceutical tests. As a result, a possible source of income for the Onge was lost, and along with it the prospect of acquiring an effective new weapon in the fight against the plague of malaria.

Origanum vulgare
(oregano)

This kind of conflict regarding the use of biological resources should really no longer occur. International law already regulates such situations. At the 1992 United Nations conference on the environment and development held in Rio de Janeiro, governments approved the Convention on Biological Diversity. It has been in effect since December 1993. In it, the signatory nations made three ambitious and interdependent commitments:
– To protect the diversity of nature – the diversity of its

Article 10
Each signatory to the accord will …
a) Include concerns for the maintenance and sustainable use of biological resources in intragovernmental decision-making processes …
c) Protect and promote customary uses of biological resources in accord with traditional cultural procedures that are compatible with the demands of preservation or sustainable use …

species and the genetic material they contain as well as the ecosystems in which they appear;

– To ensure that the use of nature is sustainable, that is, environment-friendly and capable of being continued over the long term;

– To ensure that the nations from which genetic resources come receive a "fair and just" share of any profits from the use of these resources.

In theory, the convention also granted nations that have a great wealth of species – chiefly developing countries in the tropics – new power. They have something that companies and consumers in the industrialized countries increasingly desire, but which, from now on, is under their control alone. The time is past when all living creatures were seen as "the heritage of mankind" – with the paradoxical result that it was primarily prosperous and technologically superior firms that made money by exploiting the diversity of species. Now the source countries must also receive their share. They assume the responsibility not only for protecting their natural resources, but also for keeping them accessible. However, it remains possible to patent genetic resources in the framework of internationally valid agreements.

Harpagophytum procumbens (grapple plant)

However, conflicts arise over the details of implementation: how the source countries are to share in the profits, whether in the form of technology transfers or in cash, and precisely who should receive a share of the profits. Specific answers to such questions are being discussed, but so far that is about all.

The question of how the profits are to be shared is a subject of especially hot dispute. Developing countries rich in species diversity threaten to close their borders and forbid foreign scientists to enter the country so long as the amount of money they are to receive for the use of their animal and plant worlds has not been unambiguously determined. On the other hand, companies based in the industrialized countries refuse to make binding commitments regarding what they will do in return for access, because no one knows how much money can be made by using natural materials.

There is also the issue of "intellectual property." India, for instance, sees the patenting of products made from indigenous plants and animals as an expropriation of its resources. European and American companies, however, are not willing, under any circumstances, to forgo patent rights that guarantee them exclusive rights to use products that they have developed at their own costs. Hence the United States Senate has so far declined to ratify the convention, in order to avoid possible disadvantages for American industries.

Still another problem makes it hard to declare biological resources as national property: Plants and animals do not stop at national borders. Most species' habitats include vast bioregions such as rivers, deserts, or rain forests. Only islands like Madagascar can claim an exclusive right to their species.

Although there are no clear guidelines, there are models of successful cooperation that provide something for everyone concerned. The most famous and longest running is Costa Rica's INBio, for which Felipe works. This private public service institute has made a name for itself because it is trying to draw up a systematic list of the country's estimated 500,000 species and make their potential usable without destroying them. Extracts from leaves or insects are sold to companies that want to research them in order to produce new drugs or insecticides. If they succeed in doing so, they can patent the end products. In return, the companies agree to give their contractual partner, INBio, a share of the proceeds. This kind of arrangement can mean billions of dollars in the future. INBio has also promised Costa Rican conservation officials that it will contribute half its share in the profits to the country's national parks. The rest goes for its own research work.

Arnica montana
(Arnica)

In the long run, the institute does not want to play only the role of errand boy, but wants to be active in pharmaceutical research itself. According to the contract with Merck, collaborators are to be trained in the United States and sent to staff laboratories in Costa Rica. It is considerably more lucrative to sell samples that are known to have active ingredients in them or whose chemical structures are already known, than to supply mere raw material.

Article 19
Each signatory [will], in the interest of balance and fairness, take all feasible steps to promote and facilitate access by other signatories, in particular developing countries, to the results and advantages proceeding from biotechnologies that are based on the genetic resources made available by the signatory countries.

Very promising projects are under way elsewhere, as well. In Ecua-dor's Amazon area, the California firm Shaman Pharmaceuticals is working with tribal medicine men, and in return, the firm's own foundation is helping to build schools and provide modern medical care for the natives. On the small South Atlantic island of Tristan da Cunha, Boehringer Ingelheim scientists are doing research on the blood of the natives, hoping to find a genetic key for treating asthma. Should they succeed, the people on the island who suffer from asthma will be provided with the antiasthma drug, free of charge and in perpetuity.

Professor Gerhard Seibert, the director of research on active agents at Hoechst Marion Roussel, speaks of the "high potential" of medically valuable agents in the rain forest. However, he considers the idea that billions of dollars will flow from North to South from now on an exaggeration: "The tropical countries in particular have probably overestimated the possible profits." Ultimately, finding a pharmaceutically valuable substance among 10,000 samples is a long, difficult, and expensive process. And it is very hard to tell whether the drug developed from this substance will turn out to be a best-seller.

In Gabon, Hoechst Marion Roussel is looking for pharmaceutically interesting substances in the "Forêt des abeilles" ("Forest of bees"). The local partner is the international conservation organization Pro Natura. This organization is active in twenty-five tropical countries, taking as its chief task the implementation of the social aspects of the Convention on Biological Diversity, according to which profits from the use of natural resources are to be fairly distributed between the source countries and industrial developers. The proceeds from the sale of samples and from royalties go into the Biodivalor fund. Money from this fund, which was set up in 1998, is used to pay for conservation projects in Gabon. The goal is to make sustainable use of the rain forests and to involve the local population. This is one of the many forms fair cooperation can take.

However, while INBio is seen as a model in matters of biological prospecting, its example is not often followed. Conditions in Costa Rica are particularly favorable for this sort of thing. Kenya, Mexico, and Indonesia made similar attempts, but they failed because of adverse conditions. In contrast to these other lands, Costa Rica is a free country, which gave up having its own army decades ago. The level of education is high, corruption is rare, and conservation was already highly regarded in the country before INBio was established. National parks and other reserves cover about one-fourth of the land area. Moreover, there is no doubt about who receives a share of the profits. In Costa Rica there are no indigenous peoples whose knowledge INBio could appropriate. So the difficult question does not arise, as it does elsewhere, as to who should share in the income – the chief, the medicine man, or the whole tribe? If so, and if government institutions come away empty-handed, what happens to conservation services? What happens to local scientists who are suffering from a lack of funding?

Above all, INBio is evidence of Costa Rica's exceptional success. Where nature is being heedlessly destroyed, where no scientific interlocutors exist, where property relationships are unclear – under such circumstances the search for natural substances and conservation are much less often in harmony.

In addition, many politicians and conservationists have completely unrealistic expectations as far as financial sharing goes. Even if modern technologies make it possible to find valuable substances with great speed, pharmaceutical research is still like looking for a needle in a haystack. The rule of thumb is that only one of every 10,000 samples will contain something that may later become a salable product. It is not just a matter of finding a substance that works, but also of finding one that produces no undesirable side effects.

Between 1960 and 1981, the American Institute for Cancer Research investigated about 30,000 different plants that contained some 114,000 substances, in order to find new candidates for the treatment of tumors. Yet only five substances were ultimately selected for clinical tests, and so far only one of these has been approved for medical use: taxol, which took more than twenty years to make it to pharmacy shelves. It took more than twenty years for the substance derived from the bark of the yew tree to make it to the shelves in the pharmacy.

Nonetheless, many researchers are betting on natural substances. Recently they have been concentrating on a new source of materials: the oceans. Because of their high degree of biological diversity, coral reefs have been called "the rain forests of the ocean"; in them live countless organisms that are of potential interest to pharmaceutical researchers:

- Moss animals (bryozoans) are microscopic organisms belonging to the tentaculata family. In them, scientists have been able to isolate a substance known as Bryostatin 1, with which they hope to be able to fight cancer. It inhibits the growth of leukemia cells and activates the immune system's killer cells. Bryostatin 1 is currently undergoing clinical trials.
- The Australian sponge *Cymbastela hooperi* contains a series of substances that kill off agents of malaria. However, these sponges produce the medically effective substance in such small quantities that they would be profitable only if artificially cultivated.
- Algae, for example the seaweed species *kombu* and *nori*, contain polysaccharide sulfate, which prevents infections like influenza and herpes. Scientists also attribute Japan's low rate of breast cancer to a high rate of seaweed consumption.

Searching for natural substances in coral reefs is particularly promising because they are so heavily populated that organisms have to fight to survive. Often the battle is fought with chemical weapons, a panoply of poisons and antidotes. American and Japanese researchers in particular are investigating such substances in the hope that they may be useful for maintaining human health.

Digging for green gold can result in benefits to everyone involved – if fair procedures are observed. In Namibia, with a little more money for his grapple plant roots, Fritz could finally buy a couple of goats and send his children to school. In Abrotsi, Jules could go on collecting his sweet fruits without it ever occurring to him to cut down or burn over the village forest for lack of another source of income. Chattopadhyay could be allowed to pursue every means of moving toward such an important antimalaria drug, which would at the same time make the Onge famous and perhaps wealthy. And Felipe could continue to tell visitors to Costa Rica's national parks a wonderful story about rain forests that have to be protected, not only because they are beautiful, but also because the drugs from the jungle can help millions of people who are ill.

See also plates:

Genetically Altered Plants
Page 154

The Green Pharmacy
Page 156

Nature as a Productive Force
Page 158

Meadow flowers from central Europe

Preserving Diversity: The Next Fifty Years

Over the past century, international conservation organizations such as IUCN and WWF have made great progress. Yet whether we will succeed in saving the planet's biological diversity will be decided in the next fifty years. What new strategies are there for achieving this goal?

In the beginning, there were two walls. Behind them, in the sixth century B.C.E., the ancient Persians created the first natural park protected by human beings. The two walls divided this "paradise" into two completely distinct areas. The outer wall enclosed a large wild animal preserve, in which deer, wild asses, antelopes, and wild boar roamed about, protected from poachers, but easy prey for the upper classes seeking to amuse themselves with hunting.

The second wall within the park enclosed the "paradise garden." In this refuge, gardeners and early landscape architects had largely tamed nature. Cypress groves provided shade, flower borders surrounded grassy areas, and water bubbled down artificial brooks. Here, protected from sun and wind, human beings sought quiet pleasures. Strollers walked through the park, people sat and thought, read and wrote, and whiled away the time chatting and flirting.

The double function of the Persian paradise clearly reflects the twofold relationship that has linked humans to nature over the past twenty-five centuries. The Persians and their ancestors had cut down most of the forests on the Iranian high plateau. They tried to fence off the remaining wild areas in reserves and to keep them as wild animal parks or to domesticate them in the form of gardens. This tradition continued into recent times, in which the privileged strata of European society protected their hunting ranges by inflicting draconian punishments on poachers, and amused themselves in hunting lodges or walked about in baroque gardens.

A broader movement, which emphasized not the use of nature but its aesthetic value, emerged with the nineteenth-century romantics. However, romantic painting and writing chiefly depicted poetic rural landscapes shaped by human hands, rather than untouched nature. At that time European forests had already been largely cleared and plowed. "Europe looked like a garden," complained the writer and social critic Aldous Huxley, writing about "romantic nature worship." It is easy, he noted, "to revere a weak and already conquered foe."

The first attempt to protect true wilderness was undertaken on a continent where humans had overexploited nature on an unprecedented scale: North America. In less than a century, white settlers had spread across the land as far as the Pacific, plundering its resources and laying waste to large areas. Toward the end of the nineteenth century, the first American conservation movement was formed. Its representatives worked to preserve nature as it was before the settlers arrived, and tried to establish bastions of wilderness against the advance of civilization – and thus to save at least something of the pioneer myth for later generations. In 1872 Yellowstone National Park was founded, and in 1890 Yosemite National Park in California was established.

The American example was to become the model for conservation in other parts of the world. In 1926 Krüger National Park was established in South Africa, and in 1931 Japan launched its national park system. However, conservation has only recently become a truly global political issue. A crucial step forward was made at the United Nations Conference on the Environment and Development held in Rio de Janeiro in 1992. The Convention on Biological Diversity signed on that occasion has since been ratified by most countries.

Conservation has long ceased to be a concern solely for wealthy countries. In the framework of its International Environmental Monitor, the Canadian public opinion institute, Environics, polled people in thirty countries regarding their attitudes toward conservation. The results show that concern about the environment and nature is now felt in threshold countries such as India, China, Mexico, and Argentina, and that people are increasingly demanding improvements. The insight that the preservation of nature is urgently necessary for human survival has become widespread in all parts of the world; a consensus is slowly but surely emerging. Putting this idea into practice, nevertheless, remains extremely difficult. Yet, since the growing world population is invading the last remaining empty spaces on the globe, and untouched natural land is disappearing at a rapid rate, so many species are endangered that scientists are now warning that a sixth great wave of extinction may be under way.

At the same time, all over the world, steps are being taken to protect nature. In addition to governments and international organizations such as the United Nations and the World Bank, countless nongovernmental organizations are involved in conservation efforts. New modes of action, such as those of The Nature Conservancy in the United States, avoid the bureaucratic process by buying up endangered lands with private money and establishing them as wildlife reserves. With The Nature Conservancy's help, some 280,000 square kilometers have been protected throughout the world. Promoting conservation has become a source of prestige for wealthy private individuals. Just as once the Rockefellers and Carnegies established universities and libraries, now America's superrich spend their money on nature. For instance, the multimillionaire Douglas Tompkins, who made his money in the clothing business, bought an area of 2,600 square kilometers in the forests of Patagonia, with the intention of having it run by the Chilean government as a national park.

Still more encouraging are the efforts the poorest countries are making to protect their natural resources. The small South American country of Surinam, for instance, has put one and a half million hectares of its tropical rain forests – a tenth of the country's total

Gray whale At the beginning of the twentieth century, only 2,000 specimens remained. After hunting this species was banned, the population rose, reaching 22,000 by the end of the 1990s.

land area – under protection, resisting the temptation to accept Asian timber companies' offers of large sums of money for the right to log the forests. Two further examples: The Republic of Georgia is planning to protect 20 percent of its land area, Mongolia 30 percent.

Since the 1960s, the number of protected areas has grown immensely, and it is claimed that there are now 12,413 reserves in the world each with a surface area of more than ten square kilometers (not to mention many smaller reserves). Nonetheless, the prognosis for biodiversity remains very much in doubt. The number of protected areas given may be correct, but even if it is, it reflects only half the truth. Many of these areas are still imperiled because their boundaries exist only on maps. Often it has not been possible to keep settlers, prospectors for raw materials, and poachers out of the parks. When protected areas are established, poorer countries, in particular, often lack the means to manage them.

But even so, we can maintain our resources worldwide with the funds currently available. "The impediment here is the lack of a political will to change the structure of governmental expenditures," says a team of economists and biologists at the universities of Cambridge and Sheffield. In a sensational article published in the renowned scientific journal *Nature* in 1999, they noted that, worldwide, between $950 billion and $1,450 billion are spent annually on subsidies that not only burden the public budget, but also harm nature. These "perverse subsidies" go, for example, to agriculture, fishing, and unnecessary transport, and keep prices artificially low – beneath actual market levels. With only about a quarter of the money spent on these subsidies, the British researchers maintain, the planet's biodiversity could be stabilized by establishing larger and more effectively protected reserves.

A rule of thumb used by ecologists goes like this: If the area of a protected space increases tenfold, the number of different species that are thereby protected doubles. Yet in the world today, there is frequently too little space and money to establish the desired reserves. Conservationists often find that the only alternative is to increase the protective capacity of existing reserves – for example, by setting up buffer zones that shelter protected areas from outside

	Men	Women
1.	Eagle	Cat
2.	Tiger	Butterfly
3.	Lion	Swan
4.	Dolphin	Swallow

Longing to be wild
Asked which animal they would like to be, German men and women give very different answers. Such feelings about animals are reflected in expenditures for conservation.[1]

threats or by establishing corridors connecting small bits of land with larger protected areas. Another possibility is the common management of large areas rich in biodiversity as so-called biosphere reserves. These include protected areas, along with areas used by humans. For example, among the regions of this kind listed by UNESCO are many tide flats on the North Sea coast in Denmark, Germany, and the Netherlands, the Serengeti region of Tanzania, and Australia's Great Barrier Reef.

In selecting new areas for protection, scientists now seek, more than they did earlier, to determine focal points. Since the funds available are limited, they must be focused on specific goals. Thus, areas that have high degrees of biodiversity are given priority. The American environmental organization Conservation International has drawn up a list of twenty-four such hot spots of biodiversity around the world. These contain a high number of so-called endemic species of plants and animals – that is, species that are found in these areas and in no others. Among the unquestioned hot spots are Madagascar, the Atlantic coast of Brazil, and northwest Borneo. The World Wide Fund for Nature (WWF) has undertaken to provide a still more comprehensive list of areas that are in urgent need of help. Its "Global 2000" list includes 232 regions that represent all the types of habitats on Earth. These are regions that not only have a high degree of biodiversity, but have an unusual developmental history. For example, some areas have particularly ancient species ("living fossils") or unique ecological dynamics.

In the past, often years have gone by before the plant and animal communities of an area were thoroughly inventoried. In the interim, precious time was lost in which more animals and plants died out. According to estimates made by Edward O. Wilson, a leading biodiversity expert who teaches at Harvard, as many as three species become extinct with every passing hour. Wilson's estimate is at the high end of the range, which is naturally quite broad. Estimates differ greatly because they are concerned with numbers extrapolated from the loss of rain forests and the still unknown small life-forms, such as beetles, that are assumed to live in them. Of course there is no question about the fact that we have to act as

Leopard
Since the sale of the pelts of these animals has been forbidden, the leopard's numbers in Africa have risen.

quickly as possible in order to halt the rapid disappearance of species. Toward this end, in the early 1990s, Conservation International initiated "rapid assessment programs." These involved teams of experienced species experts that could be quickly sent to endangered areas. Within a month of nonstop work, these teams can produce an inventory of the animal and plant world and inform the corresponding governments or conservation organizations of their findings. A task force composed of botanists and zoologists who participated in the first "rapid assessment program" in the upper Amazon region of Bolivia achieved spectacular success. Their discoveries, which included species of birds that are found nowhere else in Bolivia – and in one case no-where else in South America – were so impressive that in 1995 the Bolivian government, with the help of the World Bank, established the Madidi National Park on the eastern slope of the Andes.

Protecting the largest possible areas may be the primary – and most sensible – goal of conservation. Yet in practice it is often individual endangered species that attract public attention and mobilize people. Many scientists criticize this preference for "charismatic megafauna," because the biggest and most beautiful animals are not necessarily those that play the most important role in an ecosystem. However, in practice, the charming, cuddly animals do outstanding public-relations work for their endangered habitats and their co-inhabitants. A panda or a cute baby elephant that displays its charm in a zoo, a film, or an illustrated book wins hearts – and contributions – for conservation. The most charismatic animals are often also the most demanding. They need large, intact areas as their habitat. When they are protected, so are many other species that share their habitat. The panda is an example of such an "umbrella species," which spreads its protective umbrella over other species sharing its habitat. A single panda requires an area of between four and twelve square kilometers. In order to protect pandas, China has created bamboo forest reserves with a total area of more than 6,000 square kilometers. To protect tigers, sixteen reserves with a total area of 25,000 square kilometers were established. And in the United States, steps taken to protect the threatened spotted owls – each

Lion tamarin
At the end of the 1990s, 800 of these primates were living in Brazil again. Once there were fewer than 100. Breeding programs in several zoos saved the species.

211

pair of which needs 2,600 hectares of forest – halted the destruction of the last old-growth coastal rain forest in the Pacific Northwest.

Smaller and less conspicuous species thus benefit as well. The World Conservation Union's Red List of endangered species, first published in 1966, is regularly updated and currently includes 8,000 animal species and 34,000 higher orders of plants. The Convention on International Trade in Endangered Species (CITES), signed in Washington in 1973, prohibits international trade in endangered life-forms – or in products made from them. Each of the eighty signatory governments planned further steps of their own to protect species and their habitats.

It does not suffice, however, to simply establish protected areas; they must be managed in such a way that the greatest possible biodiversity is maintained over the long term. Scientists have been debating the most effective way of achieving this goal ever since the first national parks were created. Central to this debate is the question as to what extent nature can be left to take care of itself and to what extent humans should intervene to protect it. Good will often leads to unfortunate results. This is shown by the example of the oldest existing American national park, Yellowstone, which was founded over a century ago. For decades, park administrators carried out a well-intentioned program of protection that brought the area's ecosystem to the brink of collapse and led to the loss of many species. In order to protect elk, moose, and bison, the Yellowstone wolf was completely exterminated. Without natural enemies, the herds grew so large that the park was overgrazed and the landscape changed. In addition, zealous park rangers put out every fire, even if it had been started by lightning. Without the renewing effects of wildfires, a mosaic of differentiated landscapes was ultimately transformed into a monotonous biotope that was poorer in species. When a huge wildfire recently raged through the park, it gave Yellowstone a new lease on life.

The Yellowstone experience teaches us that a natural reserve may be severely impaired when people think they can control its ecological processes. We now have new strategies for managing nature that allow the ecosystem largely to take care of itself. Natural

Northern elephant seal
Only a few dozen of these seals survived hunting in the nineteenth century. A hundred years later, their numbers had grown to 150,000.

wildfires are no longer put out in Yellowstone. A decisive step back toward nature was taken with the reintroduction of wolves in the mid-1990s. This predator has proven to be a key species in the ecosystem, and contributes more to Yellowstone's health than all the human management since the park was created (see also p. 125). Even if the lesson of Yellowstone cannot be applied to all other reserves – particularly the smaller ones – a hands-off type of management generally remains appropriate. Over time, ecological changes occur in every natural system, and even in protected areas every effort should be made not to impede this process.

On the other hand, humans intervene very strongly in nature when they take endangered species out of the wild in order to breed them in zoos, aquariums, or botanical gardens. This is an expensive undertaking; keeping a rhinoceros in a zoo costs at least three times as much as protecting it in the wild. Yet these arks built by people serve as the last safety net for plants and animals whose habitats have dangerously shrunk or of which only a few individual specimens remain.

Today, species-preservation breeding is one of the most important tasks of zoos and botanical gardens. Its long-term goal is to increase the population of a species that is seriously endangered, or even already extinct in the wild, to a level that allows it to be reintroduced into its native habitat.

Since in such cases only a few members of a species still exist, zoos set up cooperative breeding programs. Information regarding the animals' genetic backgrounds is entered into data banks, and computers now often determine the optimal partner for each individual. In this way, the danger of inbreeding is diminished and the genetic diversity of the offspring maximized. Over the past few decades, species-preservation breeding has in fact saved several animals from extinction. One spectacular success is the return of the California condor. In the mid-1980s, its population had fallen to fewer than thirty individuals living in the wild, and conservationists feared it might soon become extinct. The remaining condors were caught and bred in zoos. Today, fifty of these imposing birds soar freely over California, and ninety-one more live in zoos.

Hawaiian goose
In 1948, there were only thirty birds of this species living in the wild. Since then, breeding programs in zoos have made it possible to release several thousand geese on the Hawaiian islands.

213

Przewalski's horse once again gallops through the nature reserves on the Mongolian steppe. The survival of the Arabian oryx antelope, which was extinct in the wild by the early 1960s, also seemed threatened. In the framework of a WWF project, the animals in zoos in Europe and the United States were bred and released in the deserts of Oman. Their population at first rapidly increased, though by the end of the 1990s poachers were hunting the oryx again.

The antelopes' fate clearly illustrates the dilemma of conservation: Breeding programs cannot achieve their goals if the animals' natural habitats are so severely damaged that they can no longer offer sufficient protection for the species still being born there. In any case, breeding programs can provide a solution only in exceptional cases, as we can easily see if we consider that all the zoos in the world would easily fit into the tiny country of Liechtenstein. Since every species has to be represented by a number of individuals sufficient to maintain a healthy level of genetic diversity, all the world's zoos taken together could not provide long-term shelter for more than 1,000 to 2,000 endangered vertebrate species. In addition, new breeding technologies such as cloning, which are not yet perfected, are likely to be of little benefit for the maintenance of overall biodiversity. These procedures can probably only help save from ultimate extinction a few "charismatic" species of which humans are especially fond.

A far more important contribution to species preservation is currently being made by another aspect of high technology – computerized means of communication. The Internet and data banks facilitate action by environmentalists and conservationists on nature's behalf, and they also make this action more efficient and thus less expensive and time-consuming. The Internet allows people who want to work to protect orchids, tigers, or tapirs to communicate with others all over the world who share their interests. Free access to the information stored in data banks makes existing knowledge immediately available everywhere. One major step in this direction is the Systematics Agenda 2000 project to establish a collective archive of 200 data banks, which would contain information on all of the 1.75 million plant and animal species that have been scientifi-cally described. This archive will allow anyone to page through this "catalog of life" by clicking a mouse.

Whether they are engaged in high technology or traditional con-servation, experts agree on one point: The next fifty years will be decisive for biodiversity. By 2050 the world's population will have grown by another billion people, and then remain at this high level before hypothetically decreasing again about a century later. The maximum spread of the human population will result in a bottle-neck for many plant and animal species, and we must try to save as many of our fellow life-forms as we can. Edward O. Wilson has ex-pressed his conviction that even if all possible means are immedi-ately utilized, 10 percent of current biodiversity will disappear. If we let things go, the loss will be at least 25 percent. Yet every lost life-form, no matter how small, is one too many. The philosopher Arthur Schopenhauer once put it this way: "Any fool kid can step on a beetle. But all the professors in the world can't make one."

See also plates:

Human Favorites
Page 18

**Protected Areas: Four
Countries Compared** Page 36
Survival in Zoos Page 38

Sources for text in margins:

1 Institut für Demoskopie,
    Allensbach (Germany), 1998

Maxing Out: How Many People Can the Earth Support?

Providing food, water, and space for a growing humanity is an increasingly difficult task.
Science cannot tell us at precisely what point
the earth will no longer be able to support
an increasing population – but we shouldn't
try to find out by actually reaching it.

Isaac Asimov, a futurologist and science fiction writer, reformulated
the question of the planet's capacity this way: "How many people is
the earth able to sustain? The question is incomplete as it stands.
One must modify the question by asking further: At what level of
technology? And modify it still further by asking: At what level of
human dignity?" The capacity of the earth is not a fixed magnitude,
but instead depends on the values people choose. Furthermore, it is
determined by circumstances that are constantly changing. Related
questions immediately arise: How many people with what standard
of living and with what kind of distribution of wealth? How many
people with what risk of catastrophe? How many people for how
long? How many people under what climatic conditions? How
many people in what kind of natural environment and with how
many fellow creatures?

Optimists and pessimists, scientists and philosophers, specialists
at conferences and summit meetings have all racked their brains regarding the limits of the planet's capacity – not to mention the
world's religions, which have also sought answers to this question.

Scientists have analyzed all the potentially limiting factors or observed them as a whole: energy and water use; overexploitation of the soil; pollution of the atmosphere; the food supply; and the organizational collapse of megacities. Whole brain trusts have projected the "limits of growth" or simulated them using computers.

In his book, *How Many People Can the Earth Support?*, Joel E. Cohen of New York's Rockefeller University has reviewed the most important studies from the past as well as the present. The estimates of the earth's maximum capacity range between 1 billion people to more than 1,000 billion. If we disregard the extremes, research suggests a middle ground of between 7.7 billion and 12 billion people (this corresponds, whether or not by accident, to the range of the various United Nations population projections for the next fifty years).

Taken as a whole, however, the review of the analyses of the earth's capacity shows first of all other limits: those of science. Still, two simple facts remain. First, we all know that there are limits, but no one knows what they are. Second, it would probably not be wise for humanity to try to find out what these limits are by actually reaching them.

"It may well be that the earth can provide not only 20, but even 30 or 40 billion people with a meager diet," says the German biologist Hubert Markl. "But would that be desirable?" Unless human population growth ends – and we must hope that it does – everywhere in the world, our great-grandchildren will live under conditions that now prevail in certain poor countries. According to the World Bank, some 1.5 billion people live in poverty, and have to get along on less than one dollar a day. UNICEF estimates that in the year 2000 about 1 billion people had no opportunity of learning to read and write; two-thirds of them were women. The question is whether, with the available resources, we can make it possible for more people to live in dignity, health, and sustainable prosperity. And this must be done without destroying our fellow creatures' remaining space for biodiversity. Cultural diversity and biological diversity are intimately interwoven and interdependent in multiple ways.

How large a burden of humanity is the planet carrying at the beginning of the third millennium? What is the status quo? Let us begin with a figure: 1,850 billion tons. That is currently the weight of all the living creatures on Earth.

However, this biomass represents only three-billionths of the earth's total mass. From this point of view, the manifold life in the soil, on the land surface, in the waters, and in the air is like a razor-thin icing on a pretty massive globe. All of humanity represents a proportion of less than one part per thousand in the biomass of life (calculated in dry biomass – that is, not counting the two-thirds of water). More than 99 percent of the total biomass consists of plant material. In this perspective, human beings seem a negligible burden on the planet's capacity.

However, a few items have been left out of our calculation. All life on Earth is made possible by photosynthesis, which transforms solar energy into chemical energy with the help of green plants, algae, and photosynthesizing bacteria.

All food, fossil fuels, the current biomass, and oxygen are produced by this gigantic factory. The energy that drives it is available in huge quantities: The sun will probably shine for another 5 billion years. It directs about 1.35 kilowatts per minute on each square meter of the earth's surface. "One five-thousandth of the solar energy reaching the earth's surface would suffice to meet the direct energy needs of ten billion human beings." That is how German physicist and climate researcher Hartmut Graßl describes the theoretical potential for the direct use of the sun as an energy source for civilization.

Coal from the earth is just as much stored solar energy as is a forest or a rice paddy. The domestic animals we raise to eat also contain stored solar energy that has been transformed to provide food. Our whole economy is derived from photosynthesis, as are both industry and technology. The surplus from photosyn-

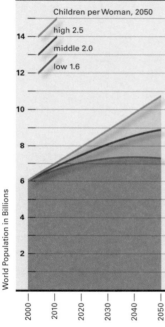

Three Scenarios of Human Population Growth
Even small declines in the number of children per woman ultimately result in large differences in the overall number of human beings. Depending on the development of this number, the population in 2050 could vary by as much as 3.5 billion. This corresponds to the total world population in 1966.

thesis ("net primary production") amounts to about 172.5 billion tons of biomass, of which about two-thirds is produced on land and one-third in the oceans.

And now comes the first surprise: Human beings, which seemed such a negligible burden on the earth's capacity to sustain life, already consume – in order to feed themselves and their domestic animals – more than one-tenth of the gigantic primary production of all the plants on land. "That is the true order of magnitude that makes us aware of the finitude of the earth," says Professor Josef H. Reichholf. This proportion cannot simply be increased, since the lion's share of the growth of biomass is produced by forests and is generally not available for human food consumption. Only a highly productive agriculture can really provide for human needs.

We share with other living creatures certain basic needs, such as food, water, and space. What is the current situation with regard to the supply and distribution of these foundations of life? So far as food goes, humans have proven to be an extremely adaptable species. Since 1930, the world population has grown from about 2 billion to over 6 billion, and yet over the same period, thanks to major advances in agriculture, the average per capita consumption of calories has increased by about 30 percent (although often it did not increase precisely where it was most needed). Countries such as India, where famines were once common, have been able to use their own resources to feed their citizens.

Even in the African Sahel, where terrible famines occurred in the 1960s and 1980s, farmers worked wonders in the 1990s. In most years the countries of the Sahel are now able to produce enough grain again. Michael Mortimore, a British specialist on Africa, concludes, "We have underestimated the ability and initiative of African farmers." If people are going hungry today, this almost always has less to do with a lack of re-

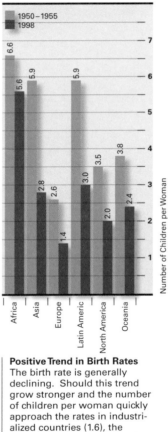

Positive Trend in Birth Rates
The birth rate is generally declining. Should this trend grow stronger and the number of children per woman quickly approach the rates in industrialized countries (1.6), the growth of the human population could come to a standstill in the coming century. 2

sources on the planet than with the fact that people are too poor to be able to buy food or grow it themselves, or because they are prevented from doing so by wars or despotism.

According to estimates based on satellite observations, about 40 percent of the world's ice free land surface is used for agriculture and two-thirds of this goes for livestock raising. Humans and their domestic animals now represent a fourth of the total animal biomass. Worldwide, humans keep more than 1.3 billion cattle, 1.8 billion sheep and goats, 900 million pigs, and 15 billion chickens, geese, ducks, and turkeys. About 40 percent of the global grain harvest is not consumed directly as food, but used as feed for livestock. And that is the bad news: It takes as many as nine kilos of grain to produce one kilo of beef. In high-yield animal production, hundreds of billions of valuable calories are burned up everyday, transformed into animal-waste heat and into huge quantities of problematic materials such as methane (a gas that makes a significant contribution to the greenhouse effect) and manure. If a cow is fed imported feed, we have to add an annual energy consumption equivalent to that of a midsized car. Herds of cattle for meat production have become major threats to wild animals and plants. Tropical forests are transformed into cattle pastures, savannas into fields producing soybeans for animal feed exported to Europe and North America. To crown it all, in 1999 alone half a million tons of government-subsidized, but unsaleable, beef had to be piled up in storehouses in the European Union.

The more than 1.3 billion cattle on Earth are not very efficient processors of feed, and they weigh as much as would 15 billion human beings. Raising fewer cattle would open up new possibilities for feeding the world's human population, and at the same time it would be one of the most important steps that could be taken toward a sustainable agriculture and the maintenance of biodiversity. As Reichholf points out, "We can make the planet of cattle the planet of human beings, in which all humans can live with dignity and in a good relationship with nature."

The proportion of the earth's surface that is devoted to farming and livestock raising cannot be further extended while at the same time maintaining biodiversity. A growing world population will necessarily have to produce its food from a smaller amount of land, per capita. Current estimates show that by 2025, an agricultural area only about forty meters by fifty meters (2,000 square meters) will be available to meet the needs of each person on Earth. In order to make crops sustainable at the higher yields necessary, better plant foods and soil care will be required. In addition, newly developed plant types that are more resistant, better adapted, and more efficient – and require less pesticide and fertilizer – will play an important role. In order to achieve this, humans have to use their most important resource: their heads. An "agriculture that pays no attention to losses" can become an "agriculture without loss" (see also the following chapter, "The Future of Life").

Our use of the oceans also leaves much to be desired. The earth is a water planet. Geologists estimate the total quantity of water on Earth at 1.4 billion cubic kilometers. It covers two-thirds of the world's surface. In terms of its volume, the habitat provided by the ocean is several times greater than that on land; if the earth's surface were completely flat, so that water was equally distributed all over the globe, the result would be a mantle of water three kilometers deep. To illustrate this, Joel E. Cohen suggests the following somewhat whimsical comparison: If all of humanity were liquid and equally distributed, the resulting film would be only half a micrometer (i.e., half a thousandth of a millimeter) deep.

The seas, with their life-forms and raw materials, still offer many opportunities for useful development. But we have not yet really understood either their manifold interactions with the atmosphere and life on land or their food chain. We may not be using their riches in an intelligent way, but rather exploiting them at a very primitive level: only two-thirds of the worldwide fish catch is used directly as food for humans. Over 20 percent is made into fishmeal for large-scale livestock raising, or processed as fish oil. About 15 percent of the catch is simply thrown overboard – the unwanted species that happen to get caught at the same time but are not marketable or suitable as food. Since the mid-1990s, 6 percent of the most important fishing grounds have been exhausted, 16 percent are being overfished, and another 44 percent are being used to full capacity. With regard to the current quantities of fish caught, the limits to what can be taken from the oceans have clearly been reached. In a completely unreasonable way, this overexploitation is still being subsidized to the tune of billions of dollars; the European Union alone spends about 1.5 billion dollars a year to protect its fishing fleets.

The productivity of nature in the water habitat is not nearly as high as it is on land. The overall population of fish and fishlike creatures does not amount to even half of the biomass of the domestic animals kept by humans. Although we can no longer draw on abundant resources, modern fishing in many respects resembles a hunt on land in which all the animals in a given part of the forest are shot dead all at once and without distinction. Only two-thirds of them would be eaten and – unimaginable as it seems – a large number of the animals would be left to rot in the forest. Doing this sort of thing on land has become unthinkable, not least of all because land is, for the most part, either in public or private ownership. One of the great challenges of the twenty-first century will be to responsibly regulate the ownership of the oceans.

Using examples from all over the world, Elinor Ostrom, a political scientist, has investigated the ways in which common goods can be protected from plundering. "Neither governments nor the market economy always succeed in creating appropriate bases for long-term, productive use of natural resources by individuals," she writes. Terry L. Anderson, an economist at Montana State University, agrees that the development of new institutions and modes of management, in governing the use of ocean resources that belong to everyone, will be possible only as the result of a long-term "evolutionary process." Accordingly, national and international institutions might grant fishing rights and other rights to cooperatives or private enterprises.

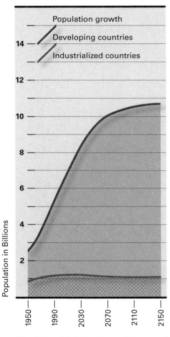

Population in Billions

Population growth
Developing countries
Industrialized countries

The New Distribution of the World's Population
Today, four-fifths of the inhabitants of the planet live in developing countries. In the future, the proportion of humanity living in these countries will be even higher. Moreover, 98 percent of the population growth occurs in these same countries.[3]

These rights would have to be defined in such a way that those holding them would have an interest in long-term maintenance of the resources and in protecting their possessions against unauthorized access. Anderson concludes, "It is time to decide who should own the oceans."

Freshwater represents only a minute proportion – about 2.5 percent – of the total water on Earth, and most of it (about two-thirds) cannot be tapped economically because it is locked up in glaciers or permanent ice, or for some other reason. Hence many scientists hypothesize that increasingly scarce supplies of drinking water will set the first limit to population growth on the planet. Brian J. Skinner, a geologist at Yale, says, "More than any other factor, the availability of water determines the maximum population capacity of a region."

The world's theoretically accessible stock of renewable freshwater is estimated at 41,000 cubic kilometers. However, a large part of the precipitation falls on the oceans, so we are unable to make use of every drop. Moreover, not all rivers can be completely directed into canals and pipes in order to be available for use. The practically useful proportion of the freshwater stock is 9,000 to 14,000 cubic kilometers.

A huge swimming pool 100 kilometers long, 90 kilometers wide, and one kilometer deep would contain 9,000 cubic kilometers. The very fact that we can so easily imagine this quantity shows how limited it is. People use over one-third of it every year. In relationship to the total theoretically available stock of renewable freshwater, this level of usage is comparable to the demands humans put on the total biomass production – humans constitute less than one tenth of a percent of the total living biomass, yet they use about 10 percent of the renewable freshwater.

According to data provided by the United Nations, the amount of water taken from rivers, lakes, reservoirs, and groundwater has grown fourfold in only

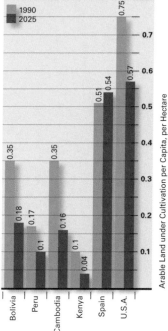

Less Land for Each Inhabitant of the Earth
The amount of arable land available for each individual is declining because of the growing world population. Almost everywhere in the world, this has to be compensated for by greater productivity if the last natural reserves are not to come under the plow. [4]

fifty years. At the same time, in many countries, polluted water is still being diverted into rivers and lakes, causing millions of people to fall ill or die. According to current predictions, by 2050 one-fourth of the world population will suffer from a lack of water. However, much could be done by using simple purification techniques and more efficient irrigation methods. Agriculture is responsible for three-fourths of present water usage, and on average less than half of this water reaches cultivated plants. If only 5 percent of the water used by farmers in the western United States were reserved for general use, cities and communities in that area would have enough drinkable water for the next twenty-five years.

However, water is not equally distributed around the globe, and its seasonal availability in many regions is difficult to predict. In water-rich countries such as Sweden, barely 3 percent of the renewable water is actually used. In European countries such as France and England – or even in Algeria or Mexico, for instance – the water usage level is between only 5 percent and 20 percent. In Libya, on the other hand, it is more than 100 percent (by deviation of underground water). Certain regions of California also use almost 100 percent of their renewable water, and in some cases more than 100 percent. If the deficit is made up by using deep-lying underground water, this only buys time – the groundwater cannot renew itself as fast as it is being pumped out.

In the Middle East, and in parts of Africa as well, there is a life-threatening lack of water. This does not alter the fact that the African continent theoretically has more available water than Europe and Asia. However, the population is not concentrated along the Congo River, which provides an abundant water supply, but rather along the Nile, for example, where three populous countries – Ethiopia, Sudan, and Egypt – are fighting over the Nile's water. Israel and Jordan squabble over the river Jordan's water, Iraq and Syria feud with Turkey over the use of the Euphrates and the Tigris. Many analysts fear that wars could result from the competition for water resources.

Thus the number of people a given region can sustain is obviously determined by certain limits, although the world as a whole may still offer enough space. Yet space must not be confused with the ability to support population. Aridity, cold, and difficult access to mountains and other regions still make about one-sixth of the world unsuited to long-term settlement by humans; on one-third they set narrow limits to the number of inhabitants.

There are still enormous uninhabited areas, such as the polar regions and some large deserts. Fewer humans and plants live in the polar regions than in deserts. Ice ages were always particularly threatening to life on land. "Of all the limiting environmental influences, cold is the most important impediment," we read in the UNESCO publication "People on Earth." But whereas cold sets the narrowest limits to vegetation and so to the basic food supply for humans and land animals, in the sea it is just the other way round: the largest concentrations of fish, seabirds, and marine mammals are found in the Antarctic and in the seas around the North Pole, where cold water rising from the depths is full of food.

Large countries such as Australia (two inhabitants per square kilometer) and Canada (three inhabitants per square kilometer) remain relatively sparsely settled – and even the African continent, with an average of twenty-five people per square kilometer, is not particularly densely populated. North America has about fourteen people per square kilometer, Latin America twenty-four. Europe has thirty-two people per square kilometer, and Asia, the most densely populated continent, has 111.

The average population density of all countries in the world is about forty-three people per square kilometer. If we take this overall average as a standard, then there should be enough room for everyone, and there should be no threat (not at least with regard to space) to biodiversity; the overall world population density is slightly greater than that in Tanzania, the African country that has some of the largest and most beautiful protected areas in the world. However, neither humans nor plants and animals are equally distributed, worldwide. Biodiversity is also concentrated in hot spots such as those in the tropical rain forests.

The hot spots of humanity can also be clearly located: 60 percent of humans are concentrated on 10 percent of the land surface.

According to a United Nations study, these places are not only the great urban centers of China, Europe, and the Indian subcontinent, but also occur in Japan, Taiwan, the Philippines, Java, Nigeria, and Egypt. In Egypt, for instance, almost all the population is concentrated in a narrow strip of land along the banks of the Nile and in its delta. The rest of this large country is too dry. What this means is that in the inhabited areas of Egypt there are currently about 1,500 people per square kilometer – one of the highest population densities in the world, apart from large cities.

Concentration and movement toward urban centers will continue. At present, the world population is increasing by about eighty million people per year. Ninety-eight percent of this growth occurs in developing countries, eighty percent of it in urban concentrations such as Cairo, Calcutta, Dakar, and Mexico City.

In 2025, 85 percent of the human population will probably live in the present developing countries. Age distributions and populations will be completely different. "In short, the earth is reinventing itself demographically," says Carl Haub, scientific director of the Population Reference Bureau in Washington (and scientific advisor to Life Counts). With regard to population growth in developing countries, it must not be forgotten, however, that between 1800 and 1930 the number of Europeans and European immigrants in North America grew twice as fast as did that of other peoples. The current growth rate of the African population, for instance, is lower than that of North America in the second half of the eighteenth century. Moreover, the growth in developing countries is not the result of families having more children than in the past. Instead, an individual's chances of surviving have been rapidly increased by better food supply and medical care. The situation of people in poor countries has definitely improved, even if much still remains to be done.

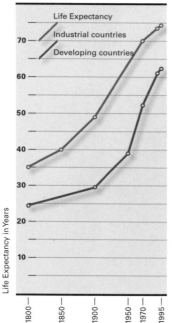

The Health Revolution
The life expectancy of people around the world has sharply increased over the last two hundred years. This is the result of progress in fighting diseases that were previously fatal. Improved overall hygiene and nutrition have also played an important role. 5

The increase in life expectancy took longer to achieve in the industrialized countries than in contemporary developing countries, where it happened almost overnight. The chances of survival improved so quickly there that there was not enough time – as there had been in the industrialized countries – to adapt social norms. In Europe and North America, a trend toward smaller families also gradually developed – but not overnight, either. Fortunately, the number of births per woman has also begun to decrease in many developing countries.

How will the world population develop in the future? That depends in a very crucial way on the choices made by the 2 billion young people in developing countries who are now under twenty years of age. How many children will they want to have, and at what age?

In the early 1950s, the average woman in the third world still raised 6.1 children, whereas today the rate has fallen below three children – lower than it has ever been before. In the more prosperous and educated Indian states, Kerala and Tamil Nadu, the average number of children per woman is now between 1.8 and 2.1. "However, the future history of Indian demography will be written in the less developed regions such as Uttar Pradesh," says Haub. "In 1998, 150 million people lived in that region and, on average, each woman had 4.8 children."

Birthrates in developing countries would decrease by about another 20 percent if the demand for family planning were fully met. Birthrates depend not only on the choices made by young people with regard to the number of children they will have, but also on whether they are able to make such choices. Today, more than 300 million couples still lack access to adequate family planning services. More than 120 million women would use birth control if they had an opportunity to do so. Surveys have shown that many women have more children than they wanted to have.

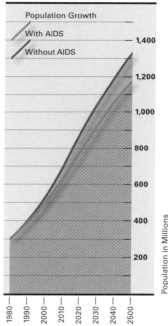

Population Growth

With AIDS

Without AIDS

1,400

1,200

1,000

800

600

400

200

1980 1990 2000 2010 2020 2030 2040 2500

Population in Millions

Decline in Africa as a Result of New Diseases
AIDS is threatening the successes achieved by improved health care in some countries. The chart shows what effects this illness will have on the population of the twenty-nine most affected African countries. Life expectancy is currently seven years below what it was without AIDS, and in Botswana it sank from sixty-one to forty-seven years of age. [6]

Couples who have, on average, two children per female only "replace" themselves, more or less, and the population ceases to grow over the long term. When this birthrate will become the world average, or when the world population will actually shrink again, demographers cannot currently predict. For its middle-level population prognosis, the United Nations assumes that the positive trend will continue and that by 2050 the average number of children per female will decrease to two (at the end of the 1990s, the rate in the industrialized countries was about 1.6). On this assumption, by 2050 the world population will have risen to 8.9 billion and be approaching its highest level. If the current birthrates implied continue, however, the population in 2050 will be 14.4 billion – more than double the current population. An optimistic estimate assumes a birthrate of 1.6 (as currently in the industrialized countries); on this assumption the world population will peak at only 7.3 billion and significantly decrease thereafter. Will world population growth stop in the new millennium? Will there even be a decline in the long run? Or will the number of people increase even faster than in the preceding century? "Any of these scenarios is possible," says Haub.

Using models, Haub's institute has calculated for Life Counts the number of human beings that have lived up to the year 2000. If we assume that the history of humanity began around 50,000 B.C.E., and ignore the numerous variations over this period of time, then we can say that about 106 billion people have lived on Earth. Scarcely 6 percent of them are currently alive. In view of the many thousands of years involved, this is an amazingly high percentage.

Humanity has become the most important species on the planet. In what it has achieved, as in what it consumes, it has reached far beyond itself. Yet its path does not necessarily lead to catastrophe. Wise self-restraint, inventiveness, a balanced approach, and intelligent use of Earth's resources can maintain the planet's ability to sustain life and open up new realms for humans and nature. This is what history teaches us. "We are burdensome to the world, the resources are scarcely adequate to us; and our needs straiten us and complaints are everywhere while already nature does not sustain us. Truly, pestilence and hunger and war and flood must be considered as a remedy for nations, like a pruning back of the human race becoming excessive in numbers." These words – freely translated – are taken from the Roman writer Tertullian's *De Anima*, which was written in Carthage in 200 C.E. In his time the world's population was one-twentieth of what it is today – only 300 million.

Sources for text in margins

1 UN, World Population
  Prospects, Revised 1998
2 Ibid.
3 Population Reference
  Bureau / UN
4 Population Reference
  Bureau / Population Action
  International, 1995
5 Population Reference
  Bureau / UN
6 Deutsche Stiftung Weltbevölkerung, 1999

Ears of corn

The Future of Life

Humans are playing an ever greater role
in shaping biodiversity. They are one of the
most successful participants in the game
of life, and to an increasing extent they themselves determine how long they will be
involved in it.

In 1980, the Swedish navy received a remarkable letter, in which
the forest administration of the island of Visingsö, in Lake Vätter,
informed the naval authorities that the wood which had been ordered for ship construction was now ready. The nonplussed naval
authorities rummaged around in their books and finally found the
order – which had been placed in 1829. It turned out that members
of the Swedish parliament had perceived a threat to the country's
future defense capabilities.

Oak was at that time considered the best material for shipbuilding, but oak trees take about 150 years to grow and mature. Since
oak forests were disappearing (because of construction needs), the
legislators foresaw that at the end of the next century (that is, in our
own time), there would be a shortage of this crucial material. So
they ordered that 20,000 young oak trees be planted on Visingsö
and reserved for the future use of the navy. Only the bishop of
Strängnäs opposed this project, arguing that although there would
surely still be wars 150 years hence, warships would probably no
longer be made of wood.

As far as war and the materials used in modern shipbuilding are concerned, the bishop was completely right. Warships have long been made of steel. But the Swedish legislators were also more correct than they imagined, although in another way. The island of Visingsö delights its residents and visitors by having one of the most beautiful oak forests in Sweden. Ordinary construction wood became a priceless source of wonder, relaxation, and contemplation.

James B. Carse, a professor of religion at New York University, uses such examples to illustrate "finite" and "infinite" games. A finite game ends with a victory. Football is such a game, and so, for the most part, are elections and modern business. The bishop of Strängnäs was right by the rules, since the Swedish navy could have spared itself the expense for tree plantations: no forest, no cost, 1-0, game won, end of game. In contrast, the goal of an infinite game is to continue indefinitely. This kind of game is played by creatures in tropical forests as well as in the jungle of human existence. Birth and death, love and reproduction, family and the course of generations – these constitute the game of life. "Finite players seek to control the future, whereas infinite players arrange things so the future keeps providing surprises." The members of the Swedish parliament saw to that. And this also holds true for the maintenance of biodiversity.

What will a forest, a tree, or a beetle be good for in a hundred years? In 1,000 years? In 10,000 years? How many species will there be, and what will they look like? A hundred years from now, our current hypotheses may seem as absurd as the tattered order slip from 1829. And yet there is something charming in the idea of a large corporation ordering an island with rare species to be delivered in the year 2200 – by a binding contract with deposit. Perhaps the Swedish parliament would like to do it again?

The American futurologist Richard Slaughter considers such speculative games highly productive, and recommends them to all of us: "It is an unusually moving thing to initiate a message which will not be read until long after one's death." Just thinking about this makes our house of cards, constructed of alleged certainties, collapse; what should we say to the people who will live 2,000 years from now? Will there still be humans? How and in what language should one communicate with them? On paper or on tape, on a computer disk or in a video? And to whom should we give this "message in a bottle"? Where can we find a mailbox for forwarding it to future generations? Anyone who writes a letter to the future becomes acutely aware of the ongoing development of human life and the various paths it might take.

Politics, religion, business, ideology, military conflicts, and cultural rivalries are some of the driving forces behind human history. However, technology is playing an increasing role. The first steps in evolution took about a billion years, while the development of the nervous system took "only" 100 million years. The triumphant progress of language took place in less than a million years. Civilization emerged within about 10,000 years. Five hundred and fifty years ago, printing began to change our lives, and the Industrial Revolution was completed in only two centuries. Over a few decades, information technology and the decoding of the human genome have changed our view of life. American supercomputer designer Daniel Hillis says: "We are dealing with an autocatalytic process" (i.e., a process that accelerates by itself). Today, people witness during their lifetimes changes as great as those formerly seen only by historians surveying past centuries.

Paradoxically, this rapid progress also makes it possible for us to recognize the very slow, lengthy processes of life, which continue over eons. "Climate and biodiversity are classic examples of this," says Stewart Brand, an American futurologist and essayist. "We have to become aware of this responsibility, because the consequences of our present conduct will probably unfold only centuries from now." Under the Antarctic ice masses, changes are now occurring that probably began 10,000 years ago. The kilometer-thick ice mantle insulates so well that temperature changes appear on the surface only thousands of years later, brought there by ice streams. Richard Alley, an Antarctic researcher at Pennsylvania State University, has looked into this question: "It may be that the natural rise in temperature that occurred ten thousand years ago (after the last ice age) is still working its way down into the ice." The memory of the pole and the climate-determining oceans is long and patient and must not be underestimated. The consequences of the changes humans are causing in the atmosphere and biosphere may well "sleep" for many years and later awaken. Minor sins are punished immediately, major ones, later on.

Stewart Brand considers it crucial that we also experience very long periods of time as present. Human civilization emerged over the last 10,000 years. "If we assume that each generation lasts twenty-five years, that makes four hundred generations," he says. "The pyramids were built two hundred generations ago." For Brand, this is as much a part of the present as the coming 400 generations or the next 10,000 years. He calls this "the long now."

To sharpen our sense of long-term responsibility, Brand has created, in collaboration with inventors, computer specialists, scientists, and artists, "the clock of the long now." This clock ticks at the beginning of each year, rings at the beginning of a century, and calls "cuckoo" at the beginning of a new millennium. The clock is supposed to function for 10,000 years.

However, can Brand and his collaborators really guarantee this? "If someone asks that question," Brand replies, "we have already achieved what we wanted – people are beginning to think long-term." And so the next question is obvious: What kind of social organization could assume the responsibility for this clock over 10,000 years? So far, no institution has lasted that long. Neither kingdoms nor superpowers have been able to survive for 10,000 years; wars and catastrophes have shaken humanity. And yet the course of things is not a history of decline. "In the long run, people's living conditions have always improved," says futurologist Paul Saffo, who is also involved in "the long now" project. Judging by the way humanity has stumbled ahead through catastrophes, he concludes, "In the short run the pessimists are right, in the long run the optimists are right."

Freeman J. Dyson, a well-known mathematician and physicist, explains the dilemma of the human battle for survival by referring to the various timescales in which the human species has to successfully survive. The human individual struggles within a horizon of years; families, within a horizon of decades. Tribes and nations try to maintain themselves over centuries; cultures, over millennia. As a species, humans strive for success over tens of thousands of years, but as a component of the biosphere the unit is millions of years. Every human being is the product of adaptation to these various timescales, and that is why conflicts in goals are so deeply rooted in our nature. What seems reasonable for the next twenty years can be a catastrophe over the next 200 years – and vice versa.

All scenarios for the future of human life and biodiversity have to be related to these various timescales. There is no great survival plan, only many small (and often contradictory) components. As strategies for the future, preservation and the creation of new opportunities are two sides of the same coin. Conservation on one hand, and scientific and technological progress on the other, can complement each other. Sometimes they are even the same thing. However, the day when all problems are solved will never dawn. There is no final weekend for the world, because every step forward involves new problems. What will count in the future, what do we have to reckon with? What do prominent futurologists foresee? Here are seven important components that will influence future biodiversity:

Infosphere and Biosphere
According to UNESCO's estimates, between 5,000 and 6,000 languages are spoken on Earth – and many of them in various dialects. "Each language reflects a society and shows how this society thinks, solves its problems, and understands the world around it," we read in the "Atlas of the World's Languages in Danger of Disappearing," published by UNESCO. When a language dies out, a whole intellectual world disappears. Specialists in linguistics assume that half of all languages have fewer than 10,000 speakers, a third have fewer than 1,000 speakers, and a tenth have fewer than 100 speakers.

As many as half of all currently existing languages could disappear in the near future. At the same time, the major regional languages will become stronger. A few Western languages, along with Chinese and Hindi, will dominate the future. This is a dramatic development, but it must be seen in an historical context. Two-thirds of the 15,000 to 20,000 languages that were once spoken on our planet have already died out in past centuries. Only a few, such as Latin, Ancient Greek, and Sanskrit, have been kept alive artificially. Students of culture, ethnologists, and representatives of "ecological linguistics" rightly deplore this loss of diversity. As in the case of the loss of species, the argument goes, the loss of languages results in a permanent loss of valuable diversity.

Nonetheless, the comparison between biodiversity and a wealth of languages does not go far enough, because cultural diversity is possible even within a single common language. In addition, new dialects and forms of speech are developing in the melting pot of the megalopolis. A world that is quickly growing together (in languages, as well) definitely has positive aspects. "Citizens of all countries are oriented toward universal human rights, and conservationists are calling on people to save the earth," say German ethnologists Joana Breidenbach and Ina Zukrigl. They add, "People, whether in Singapore or Austria, are already being almost forced to relativize their own standpoint by the constant (factual and medial) presence of other cultures. Whereas in the 1970s CBS news anchor Walter Cronkite could still end his broadcasts by saying, 'And that's the way it is,' in the early 1980s his successor Dan Rather concluded more realistically, 'That's part of our world tonight.'"

According to the cultural criticism produced mainly in Western nations, in the so-called third world a television set is a superficial luxury article. But Breidenbach and Zukrigl see it as also having positive effects. "Television gives women, young people, and the rural population in general their first equal access to knowledge and stories." Access to television enables people to operate in synchrony with the world's great megalopolises. Through the direct and diverse views television provides, desires are awakened that often remain unfulfilled. On the other hand, the stereotype of American prosperity is relativized when it is seen that in America, too, there are poor and rich, minorities and elites, criminality and illiteracy.

Just as in the biosphere, an infosphere common to all humans is developing all over the globe, and it will become even denser with the Internet. No technological barrier, national border, or political system can stop this phenomenon. It is surely easier to give people equal access to the Internet than to provide them with equal access to property and education. In the battle against the general problem of inequality, access to a computer network nevertheless is an important step forward. Thus the United Nations Development Programme (UNDP) is working, with the electronics firm Cisco Systems and other companies, on a model project that is supposed to enable people in even the smallest villages to have access to the Internet. This kind of access could help overcome the isolation of poor people and poor regions and countries, and it could also serve as a tool for establishing and extending social justice.

Billions of people live on Earth without property rights, not to mention property itself. Dictatorship, as well as political and private caprice, give them no chance. One of the biggest problems is rural poverty. And it is often greatest precisely where the most natural resources are located. Sheer necessity often forces the poor to plunder nature. Billions of humans are unable to act responsibly, which means thinking about the future. Yet today many people see no future for themselves. The long-term effects of the global infosphere could therefore prove to be extremely propitious for the global biosphere.

Bio-options

Organic farmers have a bumper sticker: "Live like you'll die tomorrow, and farm like you'll live forever." Organic farming in Europe and North America shows that agriculture is compatible with greater respect for nature and is not necessarily practiced at the expense of biodiversity. Giving up pesticides and preserving natural habitats, such as hedgerows and meadows, gives more plants and animals a chance than does conventional agriculture with its often completely cleared fields. Particularly in regions where sufficient agricultural land is available (and even left partly fallow), organic farming has emerged as a meaningful alternative to conventional agriculture. In addition, a growing number of consumers are willing and able to pay for more expensive organically grown bioproducts.

The biggest handicap of organic agriculture is definitely its need for space. Unless every last tree is cut down, it will not be possible to feed the growing world population using organic farming methods as they are practiced today. At present, weeds, pests, and plant diseases cause about one-third of the global harvest to be lost. Without the use of chemicals, this proportion would be significantly higher and the food supply for the world's population would no longer be guaranteed.

On the other hand, these chemicals are a source of many problems with water, soil, and biodiversity. By using new methods and technologies, conventional agriculture also has the potential to operate more efficiently and with greater care for the environment. For example, the international Food and Agriculture Organization (FAO) has reported on new ways of raising rice by watering at intervals, thus reducing water use by 25 percent. Such advances are urgently needed in Southeast Asia, India, and China. 90 percent of the worlds ricefields are cultivated by Asian farmers, but rice currently requires twice as much water as other kinds of grains.

Great hopes have also been pinned on breeding new plant varieties. In its bulletin "Population Change, Resources, and the Environment," the Population Reference Bureau analyzed the outlook for future development: "The currently used varieties can also be further optimized using conventional methods. Within the next twenty years, however, biotechnology might provide fundamentally new ways of increasing yields." Rice is particularly well-suited to gene research, and it has been studied by researchers around the world (along with the "model plant" *Arabidopsis thaliana* – thale cress). The knowledge thereby gained can also become the basis for developing new varieties. Currently, corn, rape, soybeans, cotton, barley, and tomatoes are the objects of breeding programs based on genetic engineering.

In such "engineering," an isolated gene with known properties is added to the genetic makeup of a life-form. "The transferred gene can come from the wild form of the plant, or from other plants, bacteria, or fungi," explains Christiane Nüsslein-Volhard, a German Nobel Prize winner. "However, this doesn't always work. For instance, in most cases we don't know which gene makes a plant resistant to frost or a given fungus. We know more about genes taken from bacteria or fungi, because they are easier to investigate than plants." Attempts are now being made to transfer an already-known resistance gene from bacteria to plants – with differing degrees of success. This kind of extremely expensive research is in its infancy and still has to show what it can really do. Scientists hope that this kind of plant development can:

– Help preserve water supplies. Worldwide, agriculture is the greatest consumer of water. New plants are supposed to produce the same yields with less water.
– Produce higher yields from dry and saline soils. The wild form's tolerance of drought and salinity is to be transferred to cultivated plants. This could improve yields in disadvantaged regions.
– Breed resistance to pests into plants and thereby reduce the use of toxic sprays while ensuring good yields.
– Improve the nutritional value of plants in order to provide better nutrition for consumers.
– Take the pressure off still untouched natural areas by increasing yields on already available agricultural land. This would indirectly help maintain biodiversity.

It is interesting to note that those involved in biotechnology argue just as ecologically as do their opponents. They seem to have a common goal, but the best way to achieve it is a subject of fierce debate. The critics of "green" genetic engineering draw attention to the following considerations:

– Genetically altered plants could spread uncontrollably or transfer certain characteristics to wild plants through interbreeding.
– The possible negative effects of this kind of breeding on human health, insects, and other small life-forms have not been sufficiently studied.
– In the future, small farmers in developing countries may no longer have access to the best seed, because it will be too expensive for them.
– The diversity of varieties could be further reduced as a result of concentration on a few widely planted varieties.

In time, it will be possible to evaluate on a broader scientific basis the opportunities and risks involved in genetically altered plants. If science and society arrive at a positive view of such plants, it is possible that organic farming and biotechnology will be mutually complementary, converging or even merging with each other.

Jeffrey A. McNeely, chief scientist of the IUCN, hopes that "a new generation of plants will be designed to produce their own nutrients and their own compounds to protect themselves against pests, thus radically reducing the need for pesticides and fertilizers. Instead of depending on chemistry, with its poisonous side effects, Golden Age agriculture will depend on biology, a science of renewal and recycling." And he adds, "Because more can be produced on less land, more territory is available for other species."

However, we must not expect miracles in the short run. Scientists at the Population Reference Bureau point out that "currently, only two percent of the worldwide biotechnology research is being conducted in less-developed countries." In addition, the seed industry's research is still oriented toward the needs of agriculture in the prosperous, temperate zones. New research and technologies are often justified by reference to the fact that people are starving. Because farmers in large areas of Africa cannot afford any fertilizer, new plants that absorbed more phosphate from the soil would be extremely helpful to them. The seeds of these plants must be reasonably priced and produce new seed that can be replanted (and not become infertile, so that farmers are forced to buy new seed every year). The development of such plants offers biotechnology firms a great opportunity to show how serious they really are about fighting hunger.

Techno-vegetarianism

Do animals have to die or live under deplorable conditions in order for humans to live? The signals from the various cultures and regions of the world are contradictory. On the one hand, all over the world the appetite for meat is increasing, even in countries that previously consumed little meat. On the other hand, in the industrialized countries, which consume the most meat, there is growing public resistance to raising animals on a massive scale under poor conditions and increasing concern that eating them may have harmful effects on human health. More and more people are eating partly or wholly vegetarian diets.

Countries like Sweden and Switzerland have already passed progressive laws concerning the raising of cattle, pigs, and chickens. The countries of the European Union are following them, step by step. Raising hens in laying batteries, for instance, will no longer be permitted in its present form after 2012.

In the long run, still more fundamental upheavals are in view. "As a matter of principle, humans are going to move farther down in the food chain and need less meat," McNeely says, "because more of the food necessary for a balanced diet is going to be produced by new varieties of plants." In his book *A Moment on the Earth*, the American environmental journalist Gregg Easterbrook goes a step further. "Theoretically," he says, "it is not necessary to raise animals in order to produce meat." Agronomists are working with the idea of using genetic engineering methods to produce meat directly from animal cells. "The end product would be genuine meat," Easterbrook writes, "but the cells would no longer make a detour through a living and suffering animal."

The thought of eating meat that was, so to speak, made in a test tube might frighten many people. On the other hand, this kind of "techno-vegetarianism" would have great advantages for the environment and for the maintenance of biodiversity. People's increasing appetite for meat would not be accompanied by more suffering on the part of animals, more destruction of the landscape, and more pollution of the environment by manure and methane. Unexpected alliances between proponents of new methods and ethically motivated consumers are conceivable and, to some extent, have already been realized. Hence, in Great Britain vegetarians and animal protectors prize a cheese that is literally "produced by genetic engineering and free of animal rennet." Rennet for traditional cheese production has to be taken from the stomachs of slaughtered calves and sheep.

Seen from a historical viewpoint, humans now consume countless substances that the hunting and gathering societies of early ages would have found highly suspect (consider, for instance, Chicken McNuggets). Housewives (and househusbands) in industrialized nations have long since forgotten how to kill a chicken. Industrially killing millions of animals to provide meat for humans may strike future generations as being just as barbaric as cutting the head off a chicken in a sparkling clean European or American kitchen. "Eventually the entire human species may converted to de facto vegetarianism," Easterbrook writes, "not through an ethical philosophy, but through the development of steak, chops, and sole that have nothing to do with animals."

New Symbioses

In their current forms, agriculture and the fishing industry constitute two of the greatest threats to biodiversity. The partial decoupling of agriculture from land and the fishing industry from the sea is in no sense a utopian idea. Food could conceivably be produced industrially from microbes, including algae, bacteria, and fungi. "This kind of food production would require scarcely any land, would be independent of climate, and would produce hardly any waste," says Joel E. Cohen, an American population researcher who takes such radical advances into consideration. In microorganisms (of which we now probably know only about 10 percent) scientists also see the potential for new kinds of biological processes for rehabilitating soil, the recovery of raw materials, and energy production that does less harm to the environment.

"Bacteria are our future," agrees German engineer Uwe Sonnenrein. He runs a shrimp breeding operation that uses recycled water. Bacteria purify the water and are then sold as a valuable fertilizer for agriculture. This innovative breeding method becomes even more significant because it could solve one of the most important environmental problems in tropical coastal areas: the destruction of mangroves by shrimp farmers. At present, mangroves are being cut down to make lagoons for shrimp farms, where pesticides and substances to prevent disease are used in massive quantities, so that after a few years only a contaminated wasteland remains. "We originally developed our technology for raising freshwater fish," Sonnenrein says, "but we have now had success in saltwater as well."

In Sri Lanka, a school for shrimp farmers has recently been established on this model. Today, 29 percent of fish eaten worldwide are produced by aquaculture. Experts assume that this proportion will rapidly increase.

Many of these operations will be involved in symbiotic relationships with other industries. A model example of this is provided by the small Danish town of Kalundborg, where a fish farm is part of what is so far a unique industrial symbiosis. The fish farm uses the waste heat from a coal-fired power plant. A biotechnology firm also takes advantage of the waste heat, using it to sterilize the bioreactors in which it produces insulin. The large amount of excess biomass is used by the local farmers as fertilizer. In all, four enterprises maintain thirteen symbiotic relationships, using their neighbors' excess energy or their waste products as raw materials (for example, the gypsum produced by filtering flue gas is immediately processed by a local lime burner). The industrial symbiosis saves annually 19,000 tons of oil, 300,000 tons of coal, 130,000 tons of carbon dioxide, and 1.2 million cubic meters of water. And it produces delicious fish.

Freeman J. Dyson adds another part of the symbiosis puzzle: "In order to help solar energy make a major breakthrough, we need a system that combines the advantages of solar cells with biological systems." Only plants and a few bacteria are capable of using photosynthesis to produce the energy required for biosynthesis. Dyson even dreams of genetically altered trees and plants that would transform sunlight into energy with an efficiency of 10 percent (instead of the current 1 percent). "It is conceivable that liquid fuel will be tapped directly from roots, in an ongoing process," Dyson writes. And he adds, "Such energy producers need not be planted in monotonous plantations, but can grow in very different ways wherever they are needed." On the roof of a factory, for instance.

The Urban Jungle as an Ecosystem

Butterflies, songbirds, rabbits, and hedgehogs see an industrial plant or a city in a way quite different from our own. They care nothing about aesthetics; only practical use-value counts: Is food available? Can I conceal myself from enemies? Will I be hunted? Especially in Europe and North America, many animals that no longer find a niche in areas cleared for intensive agriculture are moving into the cities and establishing themselves there. Their ability to adapt never ceases to amaze. Martens use cars as playgrounds and gnaw through the ignition wires; shrewd birds see cars' radiators as a cold buffet, picking the remains of insects out of them. Gardens, industrial wastelands, wastewater systems, and the cornices of high-rise buildings form the "concrete jungle ecosystem."

In New York, at least forty-seven rare or endangered plant species grow. Twelve pairs of peregrine falcons, which have become rare, have even taken up residence there. They clearly don't care whether they live under a rock outcropping or under the roof of 10 Wall Street. The main thing is that they are not hunted, and there is enough food available there (pigeons, for instance). Foxes, white-tailed deer, and coyotes have also conquered the city. Jackrabbits hop about among the planes taking off and landing at John F. Kennedy Airport. Not to mention the 30 million rats (four per resident) and the billions of cockroaches that are among the less agreeable permanent residents of the city.

Steward Picket, an ecologist working at the Institute of Ecosystems Studies in Millbrook, New York, is surprised by the way nature is penetrating the cities. Picket directs a large-scale research program that is investigating the city as an ecosystem. Field studies are intended to answer questions such as these: What makes a city attractive to foxes? Why are some parks overpopulated by gray squirrels? How does a lane bordered by maple trees affect the number of rodents? "All the things that happen in an ecosystem are happening here, just as they are happening in a tropical rain forest," says Liz Johnson of the New York Museum of Natural History's Metropolitan Diversity Program.

The larger the city, the larger the number of nesting bird species, says Professor Josef H. Reichholf, explaining that animals prefer large cities because they have become "natural paradises." In Berlin alone there are 141 species of nesting birds. In central Berlin, 380 species of wild plants, not planted by humans, grow. In the Munich metropolitan area, Reichholf identified 360 butterfly and moth species in a quadrangle of only 6,000 square meters. If we add small life-forms to this count, the average European city is home to 18,000 animal species. "Because humans are increasingly concentrated in cities, nature in urban areas is becoming an increasingly important factor in their welfare," says William W. Shaw of the University of Arizona.

According to United Nations estimates, by 2030 four out of five people in developed countries will be living in cities, and in less developed countries it will be three out of five. The great human migration is currently taking place not between different countries but between rural areas and cities. Urbanization raises serious problems for humans, especially in developing countries: first and foremost, poverty and a shortage of clean water.

Everything will ultimately depend on there being room for humans to survive outside the traditional labor market. People's ability to help themselves must not be underestimated. For example, 800 million people in the world practice – whether as workers or for their own consumption – forms of small-scale urban agriculture, ranging from vegetable gardening on New York rooftops to raising sheep in Cairo backyards. Collection of wastewater or recycling of organic garbage could be integrated into such self-invented structures.

Tree planting – for example, along highways or railroad tracks – increases people's quality of life. Trees filter out as much as 45 percent of pollutants such as nitrogen dioxide and sulfur dioxide, and when they are near buildings they can save as much as 10 percent of the energy used for heating or air conditioning.

On the overall ecological balance sheet the trend toward the city need not have only negative effects. In principle, a compact settlement pattern with short routes and more densely woven infrastructures is more protective of resources than a spacious and heavily overdeveloped landscape. In densely populated cities, for example, the use of automobiles constantly leads to traffic jams, and the consumption of fuel per resident is already much lower than in rural areas because more and more people prefer public transport. In the future, large cities in which residents are fortunate enough to have a suitable level of prosperity must not be sites of ecological and human devastation. The animal and plant worlds can also benefit under certain circumstances. For example, the rush to the city could relieve the pressure on rural areas and the plants and animals living there.

The Ambivalence of the Genome

On June 26, 2000, an international team of American, British, French, Japanese, and Chinese scientists finished decoding the human genome, which contains the text of life and is composed of 3 billion "letters." We can read this text and we can change it. Both of these abilities can turn the whole of life upside down.

This is how Daniel Cohen and J. Craig Venter, both leading gene researchers, describe the ambivalence of this enterprise: "At the end of the next century the Human Genome Project will probably be seen as a second Manhattan Project (the project that developed the atomic bomb), and the scientists involved in it as meddling Dr. Frankensteins – or else the mapping of the human genetic heritage will be regarded as the greatest advance made in the history of our species since the first primates walked upright."

Genetic science can revolutionize medicine and lead to a victory over countless lethal diseases, such as tuberculosis, cholera, and malaria. But where there is much light, there are also shadows. Lee M. Silver, an American gene researcher and evolutionary biologist, describes the way in which such good intentions can result in extremely problematic consequences. His central example is people's legitimate wish to have children. This wish is innate. Fertility clinics are among the fastest growing medical services all around the world. They use artificial means to help people have children who could not have them in the traditional way.

However, new methods of genetic engineering, known as "reproductive genetics," are opening up much broader possibilities – for instance, the possibility of having children who are free of certain genetic illnesses. We want this also. The next equally innate wish is obvious: We want not only healthy, but also successful, children. Parents are already investing huge sums in private schools and education for their children. The possibility of being able to endow children with genes that allow them to perform superlatively would be irresistible for many parents. And their wish could be fulfilled in the future. The step toward human self-breeding would thus be tacitly taken by caring parents who want only the best for their children.

"The problem with using molecular genetics is that there are no clear limits, no Rubicon that can be crossed only by making an explicit choice," says the German biochemist Jens Reich. Reproductive genetics will be the real touchstone of human ethics. And we will not be able to blame dark forces; the rift will run right through the middle of each individual. Reich sees only one way to deal with this question: a kind of glasnost. "Only open discussion of the details of what biologists do will make it possible to define reliable limits in concrete cases and to avoid criminal arrogance," Reich says.

The international Human Genome Project provides an image not only of ourselves, but also of the biosphere in which we live, a new framework. For the common origin of human beings and other creatures on Earth is preserved in the human genetic substance, DNA. Humans have much more in common with even the simplest life-form than we ever imagined before. The different creatures are coming closer together again. The deciphering of the genetic codes of other species, down to the simplest bacteria, is under way, and scientists say it promises to be useful in multiple ways. With faster analytical and calculating devices a Biosphere Genome Project may emerge at some time far in the future. Such a project would set as its goal the deciphering of all the genes in the biosphere.

Species preservation could benefit from such a project for two reasons. First, in order to decipher "the genome of the biosphere," all the species living on Earth would first have to be inventoried, named, described, and above all maintained. This would win new allies for biologists and species preservationists. Second, the preservation of these life-forms would no longer be seen as the hobbyhorse of a small group of idealists; instead, protecting them from destruction would become a century-long project that would be in everyone's interest. Jeffrey McNeely is certain that "just as machines and chemicals dominated the twentieth century, biology will be the driving force" of the twenty-first century.

The new genetic science will lead to many problems, but it could also provide an important impetus for developing our ecological consciousness; we are beginning to conceive the biosphere as the infrastructure of life. Humanity may be able to renew or reshape itself, but if it had to create the unique life network of the biosphere itself – for example, by making Mars habitable – it would be hopelessly overtaxed.

Jacques Monod, a French biochemist and essayist, suggests that an ancient alliance has been shattered. He means the identification of humans with powers and forces considered to be higher than nature. "The new alliance binds human beings to their natural environment," agrees Ilya Prigogine, a Belgian who won the Nobel Prize for chemistry. Nature will be coshaped and preserved by humans in accordance with this model, and humans will also be coresponsible for the result. Perhaps the greatest service a Biosphere Genome Project could perform would be the promotion of this idea.

There are historical parallels for such unexpected side effects of a new technology. In 1969, the Apollo space program produced the first color pictures of the earth. These photos of our beautiful, vulnerable planet gave the environmental movement an enormous boost, which culminated in 1970 in the first "Earth Day." Hardly anyone had foreseen this effect of space travel, and everyone was surprised. "Practically the whole environmental movement of the 1960s actively fought the American space program," recalls Stewart Brand, "with the exception of Jacques Cousteau." The French oceanographer was one of the first to recognize that modern technology and nature could enter into a meaningful alliance.

Is Anyone Out There?

Not only the human genome but also the universe is being inventoried – at least that part of the latter that we can see. Using electronic imaging, five American universities are collaborating in the Sloan Digital Sky Survey, with the goal of identifying and recording everything visible in the sky of the Northern Hemisphere (and, later on, in the sky of the Southern Hemisphere). Astronomers hope that this long-term project will lead to progress in their research. "The survey will provide information about the location and behavior of each object in the sky, just as the Human Genome Project provides information about the location and function of each individual gene," says Freeman J. Dyson.

At a meeting of the American Astronomical Society held in 1999, astronomers once again corrected the quantity of star systems; they now estimate that there are 125 billion star systems (each containing billions of solar systems) in the universe. To mathematicians and statisticians, it seems clear that, given the billions of other solar systems in our galaxy and the billions of other galaxies, there must be life out there somewhere. Astronomers and cosmologists no longer regard discussion of extraterrestrial life as a crazy idea. Exobiology, a young discipline that looks for traces of life on other celestial bodies, is considered a science as serious as many others.

In 1999 researchers from the United States, Canada, and Ukraine (after a pause of 25 years) have sent another radio message to the possible inhabitants of other planets. However, because of the enormous distances involved, an answer is not expected for at least another 120 years.

Up to now, traditional biologists have remained skeptical. Most of them believe that humans and the biosphere of our planet owe their existence to a scarcely conceivable chain of accidents. Perhaps we are alone, after all. American biologists Lynn Margulis and Dorion Sagan can nonetheless imagine that Earth will sow life in the universe like seeds. "Another hundred million years of sun-driven earthly plenitude should suffice to transfer life from our planet into the universe. Life has been expansionist from the beginning."

Life continues on – with or without us. Humanity must therefore take to heart a suggestion made by sociobiologist David Barash: "We all participate in a cosmic poker game in which the dealer has an endless supply of chips available. In this game neither we nor our genes ever really win, for we cannot cash in our chips and go home. There is no other game in town, and since it has been going on for a long time, only the best players are still at the table." Existence is what is at stake in this game; it is the only game we can play, and we must, in any case, try to play as long as we can. It is the game of life – Life Counts.

See also plates:

A Changing World
Page 24
Humans, a Career
Page 26

The World's Three Food Sources
Page 146

Genetically Altered Plants
Page 154

Part Three
Inventory 2000: Humans and Nature in Numbers

Any fool kid can step on a beetle. But all
the professors in the world can't make one.
Arthur Schopenhauer

A dappled horse.

Humans

106 Billion People Have Lived on Earth

This estimate was arrived at by a crude calculation based on models. "Modern humans" are assumed theoretically to have begun 50,000 years ago with a single couple (the number of humans who may have previously existed is hard to determine, and has been ignored, but would have little influence on the overall number). For about the last 10,000 years, there are clues to the size of the human population, life expectancy, and birthrates. On the other hand, probable declines resulting from disease or natural catastrophes could not be taken into account because of inadequate data, and the number of births could also have been higher.

Year	Population	Period	Births
50000 B.C.E.	2	–	–
8000 B.C.E.	5,000,000	50000 – 8000	1,137,789,769
1	300,000,000	8000 – 1	46,025,332,354
1200	450,000,000	1 – 1200	26,591,343,000
1650	500,000,000	1200 – 1650	12,782,002,453
1750	795,000,000	1650 – 1750	3,171,931,513
1850	1,265,000,000	1750 – 1850	4,046,931,513
1900	1,656,000,000	1850 – 1900	2,900,237,856
1950	2,516,000,000	1900 – 1950	3,390,198,215
1995	5,760,000,000	1950 – 1995	5,427,305,000
2000	6,055,000,000 [1]	1995 – 2000	675,000,000

The number of people ever born	106,147,380,169
The world population in the middle of the year 2000	6,055,000,000
How many of those ever born are alive in the year 2000	6 Percent

Source I

A Minute in the Life of Humanity in the Year 2000

[1] UN medium scenario

	Worldwide	Industrialized/	Developing Countries
Births	260	25	235
Deaths	101	23	78
Infant mortality	15.2	0.2	15
Growth	160	3	157

Source II

The Great Majority of Humans Live in Asia – and Have for a Long Time

Regional Distribution of the World Population, by Percentage[1]	Year 1800	Year 2000
Less developed countries	76	80
including:		
Asia (except Japan)	62	59
Africa	11	13
Latin America and the Caribbean	2	9
More developed countries	24	20
including:		
Europe	21	12
North America	1	5
Japan, Australia, New Zealand	3	2

Source III

Asia has 20 percent of the world's land surface and contains 60 percent of the world's population. In all, four-fifths of humanity live in less-developed countries.

1 Because the figures are rounded up and down, individual numbers may differ slightly from the total amounts.

Cultural Diversity

Over 6,000 Languages as an Expression of Cultural Diversity

About half of all languages have fewer than 10,000 speakers and are considered endangered.

Region	Number of Languages	Percent of World Total
Asia	2,034	31.2
Africa	1,995	30.6
Pacific area	1,341	20.5
America	949	14.5
Europe	209	3.2
Total	**6,528**	

Source IV

Which Languages Are Spoken by the Most People

Number of People Who Speak a Language as Their First Language, in Millions	
Mandarin (China)	885
Spanish	332
English [1]	322
Bengali (Bangladesh)	189
Hindi (India)	182
Portuguese	170
Russian	170
Japanese	125
German	98
Wu (China)	77
Javanese (Indonesia)	76
Korean	75
French	72
Vietnamese	68
Telegu (India)	66

Source V

[1] If first and second languages are taken together, English is in second place, with 470 million speakers.

Languages on the Internet

Languages	Percentage of Internet Use
English	57.4
Japanese	8.8
German	6.2
Chinese	4.4
Spanish	4.3
French	4.2
Scandinavian languages	3.3
Italian	2.5
Dutch	2.0
Korean	1.9
Portuguese	1.5
Other	3.5

Source VI

On the Internet, more than one out of two users understands English, but since the mid-1990s the number of non-English-speaking users has been slowly but steadily increasing.

Biosphere

All of Life on Earth Weighs
1,850 Billion Tons

Over 99 percent of the earth's biomass consists of plants. Humans represent a minuscule part of the total biomass. Yet they and their domestic animals consume a tenth of the total annual production of land plants.

What Life Weighs

	In Millions of Tons
Total biomass [1] of all living creatures	1,850.0
including, for instance:	
Biomass of tropical rain forests	765.0
Biomass of animal life, total	2.3
Biomass of domestic animals, chiefly cattle	0.4
Biomass of human beings	0.1

How Much Life Grows Each Year

Total annual growth (net primary production)	172.5
including:	
On continents	117.5
In the open ocean	41.5

How Productive Specific Ecosystems Are (Annual Net Primary Production)

Tropical rain forests	37.4
Monsoon forests	12.0
Tropical savannas	10.5
Northern conifer forests	9.6
Deciduous forests	8.4
Cultivated/agricultural land	9.1

Source VII

1 Dry weight

Biosphere

How Much Biomass Grows Each Year per Human Individual, and How Much Each Human Individual Uses	In Tons
Total	28.6
Land	19.6
Oceans	6.9
Cultivated land	1.5
Land-plant consumption by humans [1]	ca. 2.0
	Source VII

How Weight Is Distributed in the City

Biomass Distribution [2] Based on the Example of Brussels (Belgium)	Tons	Proportion by percentage
All plants	750,000	91.50
1,075,000 residents	59,000	7.16
Earthworms	8,000	0.97
Other animals	5,000	0.61
100,000 dogs	1,000	0.12
250,000 cats	750	0.09
		Source VIII

Earthworms weigh eight times as much as all the dogs in a city.

[1] As food for humans and animals
[2] In live weight

Humans and the Environment

For Every Human Being There Are about 500 Trees

Worldwide, the total surface in protected areas is almost as large as that under cultivation.

In the Year 2000, the Following Areas Correspond to Each Person: [1]

	In square meters
Total land surface	25,000
Forest	6,000
Agricultural land	2,500
Areas in large reserves [2]	2,300
Number of trees per person [3]	500

Quantity of Water Available for Human Use, per Capita and per Annum:

	In cubic meters
Renewable freshwater	6,833
Practically usable proportion	1,500
Actual use	645

Source IX

One Tree Produces 3 Million Liters of Oxygen per Year

An Average Deciduous Tree 15–20 Meters High:

Total leaf surface	ca. 1,000 cubic meters
Amount of organic material produced	4,000 kg. per year
Amount of oxygen produced	3 million liters per year
	370 liters per hour
Water consumption for oxygen production	2,500 liters per year
Filtering capacity of the foliage (dust, etc.)	7,000 kg. per year
Root mass	300 – 500 kg.
Water runoff prevented by roots	70,000 liters per year
Its roots run through	1 ton of humusy soil
	50 tons of rocky soil

Source X

1 Figures rounded off, adjusted to current population of 6 billion people.
2 Not including the world reserve Antarctica or reserves of less than 10 square kilometers.
3 This is the average of the number for tropical rain forests and the number for Central European deciduous forests.

Humans and the Environment

Biodiversity Can Be Saved at One-Fourth the Cost of Destructive Subsidies

Expenditures, World Average, per Capita and per Annum:	
	In U.S. dollars
Actual expenditures for conservation areas	1
Funds necessary to save biodiversity [1]	50
Government subsidies that harm nature [2]	200
Gross domestic product, worldwide	5,170
Gross domestic product, North America	28,130

Source XI

According to British scientists, biodiversity can be saved with the available means.

Most Biodiversity Has Not Yet Been Discovered

Estimated number of still unknown species	10,000,000 – 200,000,000
Number of all known species	1,750,000
Insects	950,000
Plants	270,000
Spiders	75,000
Mollusks	70,000
Crabs	40,000
Fish	25,000
Birds	9,950
Reptiles	7,400
Amphibians	4,950
Mammals	4,630

Source XII

1 Economists have calculated that an effective program to protect biodiversity on 10 percent of the world's surface would cost $300 billion.
2 For example, for agriculture, fishing, energy, water, and transportation systems. The estimates range from $950 billion to $1,450 billion. Here we have assumed a middle-level cost of $1,200 billion.

Humans and Animals

More Animals on Earth than Stars in the Galaxy

According to a prudent estimate made by experts, the roughly approximated number of individuals of all animals is calculated to be about 1 quintillion, and consists almost entirely of insects and other small life-forms. Other scientists think there may be ten times as many.

The number of all animals on the planet	1,000,000,000,000,000,000
	(one quintillion)
Ants	10,000,000,000,000,000
	(ten quadrillion)
Birds [1]	300,000,000,000
Stars in our galaxy	200,000,000,000

For each human being there are:	
Birds	50
All animals	167,000,000
Bacteria [2]	1,000,000,000,000,000,000,000
	(one sextillion)

Source XIII

For Every Elephant There Are 10,000 Human Beings

How Many Human Beings There Are for Every Animal of Various Species [3]

Animal	Human beings
Elephant	10,000
White stork	20,000
Lion	100,000
Tiger	1,000,000
Panda	5,000,000
Sumatran rhinoceros	100,000,000
Spix's macaw [4]	6,000,000,000

Source XIV

1 Middle-level estimates range from 200 billion to 400 billion.
2 An estimate of about 6 quintillion ($4-6 \times 10^{30}$) bacteria worldwide is assumed here.
3 Each midlevel value based on various population estimates.
4 There is only one living specimen of this species of parrot living in the wild.

Wild Animals

Land Mammals

Animal	Population	Trend
Chimpanzee	105,000 – 200,000	stable/sharply declining [1]
Gorilla [2]	115,000 – 122,000	stable/sharply declining [1]
Orangutan	30,000 – 50,000	declining
Golden-headed lion tamarin	550 – 600	unknown
Lion	30,000 – 100,000	declining
Tiger	4,600 – 7,200	relatively stable [3]
Wolf	118,000 – 146,000	generally declining [4]
Brown bear	185,000 – 200,000	unknown
Polar bear	22,100 – 27,000	stable
Spectacled bear	10,000	declining
Giant panda	1,200	sharply declining
Bison	over 200,000	increasing
Cape buffalo [5]	560,000 – 1,000,000	declining
White-tailed deer	19,600,000	sharply increasing
Reindeer [6]	2,900,000	unknown
Red deer [7]	2,000,000	increasing
Black rhinoceros	2,400	slightly increasing [8]
White rhinoceros	7,000	increasing [9]
Indian rhinoceros	1,800 – 2,000	stable
Sumatran rhinoceros	500	declining
Javan rhinoceros	50 – 80	stable [10]
African elephant	540,000	stable [11]
Asian elephant [12]	38,000 – 49,000	declining
Red kangaroo	9,600,000	increasing
Koala	20,000 – 80,000	slightly increasing

Source XV

There are secure global populations of only a few prominent species. This list shows the current state of our knowledge about the worldwide frequency of bird and mammal species. The WCMC produced it by combining the few scientific censuses with generally accepted estimates. See the chapter, "Statisticians on Safari," and the plate, "Human Favorites."

1 By region
2 All subspecies
3 After a major decline
4 Slight recovery in some areas
5 Including the red buffalo
6 Only wild populations, including caribou in North America
7 All subspecies, worldwide
8 After a decline of 85 percent
9 Northern subspecies (32 animals) seriously endangered
10 At an extremely low level
11 After a major decrease
12 Wild population

Wild Animals

Marine Mammals

Animal	Population	Trend
California sea lion [1]	over 175,000	sharply increasing
Gray seal [2]	167,000 – 198,000	increasing
Harp seal	2,500,000 – 4,700,000	decreasing
Southern elephant seal	700,000 – 800,000	relatively stable [3]
Walrus	240,000	unknown
Bottle-nosed dolphin (*Tursiops truncatus*)	95,000 [4]	unknown
Spinner dolphin [5]	633,000	unknown
White whale	100,000	declining
Narwhale	35,300	declining
Sperm whale	2,000,000	unknown
North cape whale	under 1,000	unknown [6]
Southern right whale	1,500 – 4,000	unknown
Bowhead whale	under 8,500	slightly increasing
Blue whale	under 5,000	unknown
Fin whale	50,000 – 100,000	unknown
Sei whale	65,000	unknown
Gray whale	22,000	increasing
Humpback whale	20,000	increasing
Minke whale	610,000 – 1,284,000	debated

Source XVI

1 Including the Galapagos subspecies
2 Only populations in the northwest Atlantic, British coast, and the Baltic
3 Increasing or declining, depending on the region
4 Number covers only the northeast coast of the United States, the Gulf of Mexico, the Pacific, the Japanese coast, and the Mediterranean. Other marine areas: unknown.
5 Only spinner dolphin
6 Seriously endangered because of the extremely small population

Wild Animals

Birds

Animal	Population	Trend
Rock dove/feral pigeons [1]	12,000,000 – 32,000,000 [2]	probably increasing
Mourning dove	475,000,000	stable
House sparrow	120,000,000 – 400,000,000 [2]	unknown
Red-billed quelea	1,500,000,000	stable
Skylark	74,000,000 – 320,000,000 [2]	declining
Swallow	13,000,000 – 33,000,000 [2]	unknown
Herring gull	2,300,000 [3]	stable
Red knot	1,200,000 – 1,350,000	stable
Curlew sandpiper	1,000,000	stable
White stork	300,000	increasing
Greater flamingo	700,000	increasing
Chilean flamingo	over 200,000	stable
Andean flamingo	50,000	declining
Eurasian crane	220,000 – 250,000	increasing
Great bustard	26,000 – 32,000	declining
Bald eagle	110,000 – 150,000	increasing
Steller's sea eagle	7,500	unknown
Philippine cockatoo	1,000 – 4,000	declining
Spix's macaw	1	nearly extinct [4]
Hyacinth macaw	3,000	declining
Jackass penguin	100,000 – 340,000	declining
Humboldt penguin	20,000	declining
Magellanic penguin	4,500,000 – 10,000,000	stable
King penguin	over 2,000,000	stable
Emperor penguin	270,000 – 350,000	stable

Source XVII

1 Rock doves and domestic pigeons that have returned to the wild
2 In Europe
3 In northern Europe and North America
4 39 more in zoos

Wild Animals

Mammals, the Example of Great Britain:
Only the Field Vole Is More Common than Humans [1]

Leading mammal specialists in Great Britain report the number of wild mammals in the country recorded in a national census. On scientific grounds, the results can be considered a realistic assessment.

Animal	Population	Trend
Field vole	75,000,000	declining
Humans [2]	59,400,000	stable
Common shrew	41,700,000	stable
Wood mouse	38,000,000	stable
Rabbit [3]	37,500,000	increasing
Mole	31,000,000	stable
Bank vole	23,000,000	stable
Pygmy shrew	8,600,000	stable
Common rat [3]	6,790,000	declining
House mouse [3]	5,192,000	declining
Gray squirrel [3]	2,520,000	increasing
Pipistrelle bat	2,000,000	declining
Water shrew	1,900,000	probably stable
Hedgehog	1,555,000	declining
Harvest mouse	1,425,000	declining
Water vole	1,169,000	declining
Brown hare [3]	817,000	declining
Feral housecats [3]	813,000	probably stable
Yellow-necked mouse	750,000	declining
Roe deer	500,000	increasing
Common dormouse	500,000	probably declining
Stoat	462,000	declining
Weasel	450,000	declining
Red deer	360,000	increasing
Mountain hare	350,000	declining
Badger	250,000	probably stable
Red fox	240,000	increasing
Brown long-eared bat	200,000	probably declining
Red squirrel	160,000	declining

1 Mammals in Great Britain, in the order of their frequency. Based on the annual minimum number of individuals of a species, before reproduction.
2 Population figures for Great Britain and Ireland
3 These mammals did not originally live in the British Isles, but were brought there by humans.

Wild Animals

Animal	Population	Trend
Daubenton's bat	150,000	probably stable
North American mink [3]	110,000	increasing
Fallow deer [3]	100,000	probably stable
Natterer's bat	100,000	probably stable
Gray seal	93,500	increasing
Noctule bat	50,000	declining
Chinese muntjac ("barking deer") [3]	40,000	increasing
Whiskered bat	40,000	probably declining
Common seal	35,000	increasing
Brandt's bat	30,000	probably declining
Polecat	15,000	increasing
Serotine bat	15,000	probably stable
Lesser horseshoe bat	14,000	probably stable
Lesser white-toothed shrew [3]	14,000	stable
Sika deer [3]	11,500	increasing
Fat dormouse [3]	10,000	increasing
Leisler's bat	10,000	stable
Otter	7,350	increasing
Barbastelle bat	5,000	probably declining
Greater horseshoe bat	4,000	declining
Pine marten	3,650	increasing
Feral goat [3]	3,565	stable
Wildcat	3,500	stable
Feral ferret [3]	2,500	stable
Feral sheep	2,100	stable
Bechstein's bat	1,500	stable
Grey long-eared bat	1,000	stable
Ship rat [3]	1,300	declining
Chinese water deer [3]	650	increasing

Source XVIII

Farm Animals

Humans Keep over 20 Billion Domestic Animals and 3 Trillion Bees

Animal	Number
Bees [1]	3,172,864,740,000
Chickens	13,478,302,000
Cattle	1,318,386,030
Sheep	1,064,110,170
Pigs	935,614,106
Ducks	773,476,000
Goats	699,994,077
Rabbits	452,345,000
Turkeys	246,462,000
Geese	209,227,000
Domestic buffalo	162,362,481
Horses	60,945,643
Donkeys	43,364,808
Camels	19,083,344
Mules [2]	14,149,019
Llamas and alpacas	5,450,000
Farm crocodiles	2,600,000

Source XIX

1 Calculated on the basis of 52,881,079 hives containing an average of 60,000 bees per hive
2 A cross between an ass and a horse

In Europe, the Most Common Pets Are Fish

Number of Pets (Western Europe and the United States Only)	
Cats (not including strays)	106,000,000
Dogs (not including strays)	94,000,000
Pets in Europe	261,000,000
Including:	
Fish	102,000,000
Cats	47,000,000
Dogs	41,000,000
Birds	35,000,000
Other	36,000,000
	Source XX

Sources: Inventory 2000

I Population Reference Bureau

II Deutsche Stiftung Weltbevölkerung, Weltbevölkerungsuhr 1999

III Population Reference Bureau, Population Bulletin, March 1999/ Deutsche Stiftung Weltbevölkerung

IV UNESCO's World Atlas of Disappearing Languages

V The Ethnologue, World Wide Web pages, Update 1999

VI Global Research, Nua

VII J. H. Reichholf, Der Blaue Planet, Munich 1998, Geographische Rundschau Nr. 51, 1999, Lexikon der Biologie, 1992

VIII D. Kalusche, Ökologie in Zahlen, Stuttgart 1996

IX J. E. Cohen, How Many People Can the Earth Support?, 1995, Bundesministerium für Ernährung, Landwirtschaft und Forsten, Deutsche Stiftung Weltbevölkerung, Stiftung Entwicklung und Frieden: Globale Trends 1998/ WCMC/ R. Irslinger, University of Rottenburg

X D. Kalusche, Ökologie in Zahlen, Stuttgart 1996

XI A. N. James et al., Nature, Vol. 401, September 1999, Deutsche Stiftung Weltbevölkerung 1999

XII WCMC

XIII–XIV J. Illies, B. Hölldobler/ E. O. Wilson, WCMC, K. J. Gaston / T. M. Blackburn, Harenberg, Kompaktlexikon 1996

XV–XVII WCMC

XVIII J. Harris, P. Morris, S. Wray, D. Yalden: A Review of British Mammals – Population estimates and conservation status of British mammals other than cetaceans, JNCC, 1995, Deutsche Stiftung Weltbevölkerung / Population Reference Bureau 1999

XIX FAO / Krokodile, D. Jelden, Bundesamt für Naturschutz

XX FEDIAF 1996, Humane Society, WWW pages, USA 1998

XXI WCMC

Biodiversity in 200 Countries

Country	Mammal species, total	Mammal species, endemic	Bird species, total	Bird species, breeding there	Bird species, endemic	Reptile species, total	Reptile species, endemic	Amphibian species, total	Amphibian species, endemic	Higher plant species, total	Higher plant species, endemic
Afghanistan	123	2	460	235	0	103	4	6	1	4000	800
Albania	68	0	306	230	0	31	0	13	0	3031	24
Algeria	92	2	375	192	1	81	4	10	0	3164	250
American Samoa	3	0	50	34	0	11	0	0	0	471	15
Andorra	44	0	147	113	0	16	0	6	0	1350	–
Angola	276	7	909	765	12	–	19	–	22	5185	1260
Anguilla	3	0	61	–	0	11	2	1	0	321	1
Antarctica	0	0	18	17	0	0	0	0	0	41	11
Antigua and Barbuda	7	0	140	49	0	13	4	2	0	845	–
Argentina	320	49	976	897	19	234	78	153	45	9372	1100
Armenia	84	3	346	242	0	51	0	7	0	–	–
Aruba	–	0	172	48	0	10	3	2	0	460	25
Australia	260	206	751	649	350	748	641	205	183	15638	14074
Austria	83	0	414	213	0	14	0	20	0	3100	35
Azerbaijan	99	0	360	248	0	54	0	10	0	4300	240
Bahamas	12	3	222	88	3	35	17	2	0	1111	118
Bahrain	17	0	294	28	0	25	0	1	0	195	–
Bangladesh	109	0	684	295	0	119	2	19	0	5000	–
Barbados	6	0	172	24	0	9	3	1	0	572	3
Belarus	74	0	298	221	0	7	0	12	0	2100	–
Belgium	58	0	429	180	0	8	0	17	0	1550	1
Belize	125	0	533	356	0	107	2	32	1	2894	150
Benin	188	0	630	307	0	–	1	–	0	2201	–
Bermuda	3	0	345	8	1	1	1	0	0	167	15
Bhutan	99	0	543	448	0	19	2	24	0	5468	75
Bolivia	316	16	1275	–	18	211	20	122	28	17367	4000
Bosnia-Herzegovina	72	0	316	218	0	27	0	16	0	–	–
Botswana	164	0	550	386	1	157	2	38	0	2151	17
Brazil	417	119	1635	1500	185	491	201	581	375	56215	–
British Territory in Indian Ocean	–	0	45	14	0	0	0	0	0	101	–
Brunei	157	0	438	359	0	44	0	76	0	6000	7
Bulgaria	81	0	374	240	0	33	0	17	0	3572	320
Burkina Faso	147	0	453	335	0	–	3	–	0	1100	–
Burundi	107	0	596	451	0	–	0	–	2	2500	–
Cambodia	123	0	429	307	0	82	1	28	0	–	–
Cameroon	409	14	1000	690	8	183	21	190	66	8260	156
Canada	193	7	578	426	5	41	0	41	0	3270	147
Cape Verde	5	0	128	38	4	12	9	0	0	774	86
Cayman Islands (British)	8	0	180	45	0	18	6	1	0	539	19
Central African Republic	209	2	662	537	1	129	0	47	0	3602	100
Chad	134	1	532	370	0	5	1	–	0	1600	–
Chile	91	16	448	296	16	82	43	49	34	5284	2698
China, People's Republic of	400	83	1244	1103	70	340	81	290	158	32200	18000
Colombia	359	34	1721	1700	67	593	115	684	230	51220	1500
Comoros	12	2	91	55	14	22	7	2	0	721	136
Congo, Democratic Republic of	450	28	1086	929	24	377	35	80	53	11007	1100
Congo, Republic of the	200	2	569	449	0	–	1	–	1	6000	1200
Cook Islands	1	0	50	27	6	–	0	0	0	284	3
Costa Rica	205	7	850	600	6	214	38	168	39	12119	950
Côte d'Ivoire	230	0	694	535	2	–	3	–	3	3660	62
Croatia	76	0	358	224	0	29	0	20	0	–	–
Cuba	31	12	342	137	21	105	83	55	50	6522	3229
Cyprus	21	1	347	79	2	23	1	4	0	1682	–
Czech Republic	81	0	398	199	0	10	0	19	0	–	–
Denmark	43	0	439	196	0	5	0	14	0	1450	1
Djibouti	61	0	359	126	1	36	0	3	0	826	6
Dominica	12	0	163	52	2	14	2	2	1	1228	11
Dominican Republic	20	0	254	136	0	117	34	35	15	5657	1800

Country	Mammal species, total	Mammal species, endemic	Bird species, total	Bird species, breeding there	Bird species, endemic	Reptile species, total	Reptile species, endemic	Amphibian species, total	Amphibian species, endemic	Higher plant species, total	Higher plant species, endemic
Ecuador	302	25	1559	1388	37	380	120	426	162	19362	4000
Egypt	98	7	439	153	0	83	0	6	0	2076	70
El Salvador	135	0	420	251	0	73	4	23	0	2911	17
Equatorial Guinea	184	1	322	273	3	–	4	–	2	3250	66
Eritrea	112	0	537	319	0	90	0	19	0	–	–
Estonia	65	0	330	213	0	5	0	11	0	1630	–
Ethiopia (before partition)	277	31	861	626	28	206	11	69	36	6603	1000
Falkland Islands	0	0	183	64	4	0	0	0	0	165	14
Faroe Islands	–	0	259	71	0	0	0	0	0	236	1
Federal States of Micronesia	6	3	104	40	18	–	2	0	0	1194	293
Fiji	4	1	109	74	24	25	11	2	2	1518	760
Finland	60	0	425	248	0	5	0	5	0	1102	–
France	93	0	506	269	1	32	1	32	3	4630	133
French S. Territ. and Antarctica	–	0	–	48	3	0	0	0	0	–	–
French Guiana	150	3	707	–	1	131	1	89	2	5625	144
French Polynesia	0	0	81	60	25	10	0	0	0	959	560
Gabon	190	3	629	466	1	–	3	–	4	6651	–
Gambia	117	0	504	280	0	47	1	30	0	974	1
Georgia	107	2	300	–	0	52	0	13	0	4350	380
Germany	76	0	503	239	0	12	0	20	0	2682	6
Ghana	222	1	725	529	0	–	1	–	4	3725	43
Gibraltar	7	0	282	34	0	15	0	10	0	600	–
Great Britain and N. Ireland	50	0	590	230	1	8	0	7	0	1623	16
Greece	95	3	398	251	0	56	8	15	2	4992	742
Greenland	9	0	–	62	0	–	0	–	0	529	15
Grenada	15	0	150	50	1	16	1	3	0	1068	4
Guadeloupe	11	4	134	52	2	20	2	5	2	1400	26
Guam	2	0	79	18	2	11	1	0	0	330	69
Guatemala	250	3	669	458	1	235	24	107	34	8681	1171
Guinea	190	1	552	409	0	–	3	–	3	3000	88
Guinea-Bissau	108	0	319	243	0	–	2	–	1	1000	12
Guyana	193	1	737	678	0	–	2	–	14	6409	–
Haiti	3	0	220	75	1	108	35	56	27	5242	1623
Honduras	173	2	684	422	1	162	22	81	34	5680	148
Hungary	83	0	361	205	0	15	0	16	0	2214	38
Iceland	11	0	316	88	0	0	0	0	0	377	1
India	316	44	1219	926	58	390	188	209	122	16000	5000
Indonesia	457	222	1531	1530	408	514	305	285	115	29375	17500
Iran	140	6	502	323	1	167	29	11	5	8000	–
Iraq	81	2	381	172	1	81	1	6	0	–	–
Ireland	25	0	417	142	0	1	0	3	0	950	–
Israel	116	4	511	180	0	97	1	7	0	2317	–
Italy	90	3	490	234	0	40	1	41	12	5599	712
Jamaica	24	2	262	113	26	36	27	24	21	3308	923
Japan	188	42	665	250	21	87	33	61	45	5565	2000
Jordan	71	0	361	141	0	73	0	–	0	2100	–
Kazakhstan	178	4	489	396	0	49	0	12	0	–	–
Kenya	359	23	1068	847	9	190	18	88	10	6506	265
Kiribati	–	0	69	26	1	–	0	0	0	60	2
Korea, North	–	0	390	115	1	19	2	14	2	2898	107
Korea, South	49	0	372	112	0	25	1	14	0	2898	224
Kuwait	21	0	321	20	0	29	0	2	0	234	–
Kyrgyzstan	83	1	368	–	0	33	0	4	0	3786	–
Laos	172	0	651	487	1	66	1	37	2	–	–
Latvia	83	0	325	217	0	7	0	13	0	1153	–
Lebanon	57	0	329	154	0	42	1	8	0	3000	–
Lesotho	33	0	281	58	0	–	2	–	0	1591	2
Liberia	193	0	581	372	1	62	2	38	4	2200	103

0 = none
– = no data

Country	Mammal species, total	Mammal species, endemic	Bird species, total	Bird species, breeding there	Bird species, endemic	Reptile species, total	Reptile species, endemic	Amphibian species, total	Amphibian species, endemic	Higher plant species, total	Higher plant species, endemic
Libya	76	5	323	91	0	56	1	3	0	1825	134
Liechtenstein	64	0	235	124	0	7	0	10	0	1410	–
Lithuania	68	0	305	202	0	7	0	13	0	1796	–
Luxembourg	55	0	289	126	0	7	0	14	0	1246	–
Macedonia	78	0	330	210	0	31	0	13	0	3500	–
Madagascar	141	93	266	202	105	363	259	179	155	9505	6500
Malawi	195	0	645	521	0	124	7	69	3	3765	49
Malaysia	300	36	736	508	18	350	71	189	70	15500	3600
Maldives	3	0	125	23	0	0	0	0	0	–	–
Mali	137	0	622	397	0	16	4	–	1	1741	11
Malta	22	0	395	26	0	8	0	1	0	914	5
Marshall Islands	0	0	75	17	0	7	0	0	0	100	5
Martinique	9	0	131	52	1	9	3	1	0	1287	30
Mauritania	61	1	541	273	0	–	1	–	0	1100	–
Mauritius	4	1	81	27	8	11	11	0	0	750	325
Mayotte	–	0	–	27	2	15	1	–	0	–	–
Mexico	491	140	1054	772	92	704	368	310	194	26071	12500
Moldova	68	0	270	177	0	9	0	13	0	1752	–
Monaco	–	0	–	–	0	6	0	3	0	–	–
Mongolia	133	0	434	426	0	22	0	6	0	2823	229
Montserrat	7	0	111	37	1	11	3	2	0	671	2
Morocco	105	4	416	210	0	90	11	11	1	3675	625
Mozambique	179	2	678	498	0	167	5	62	1	5692	219
Myanmar	251	6	999	867	4	203	37	75	10	7000	1071
Namibia	250	3	644	469	3	250	25	51	2	3174	687
Nauru	–	0	22	9	1	–	0	0	0	50	1
Nepal	181	2	844	611	2	100	1	43	11	6973	315
Netherlands	55	0	456	191	0	7	0	16	0	1221	–
Netherlands Antilles	–	0	252	77	0	18	4	2	0	–	–
New Caledonia	11	3	68	107	22	58	45	1	0	3250	3200
New Zealand	2	2	287	150	74	52	48	3	3	2382	1942
Nicaragua	200	2	750	482	0	161	6	59	2	7590	40
Niger	131	0	482	299	0	–	0	–	0	1170	–
Nigeria	274	4	862	681	2	135	7	109	1	4715	205
Niue	1	0	29	15	0	4	0	0	0	178	1
Northern Marianas	–	0	88	28	2	11	0	0	0	315	81
Norway	54	0	453	243	0	5	0	5	0	1715	1
Oman	56	2	430	107	0	64	8	–	0	1204	73
Pacific Islands (U.S.A.)	–	0	–	–	0	–	0	–	0	–	–
Pakistan	151	4	671	375	0	172	23	17	4	4950	372
Palau	2	0	135	45	10	22	2	1	1	–	–
Panama	218	16	929	732	9	226	25	164	22	9915	1222
Papua New Guinea	222	65	708	653	94	280	80	225	128	11544	–
Paraguay	305	2	600	556	0	120	3	85	3	7851	–
Peru	460	49	1710	1541	112	360	96	376	152	17144	5356
Philippines	158	102	557	196	186	190	159	92	73	8931	3500
Pitcairn Islands	0	0	26	19	5	5	0	0	0	76	14
Poland	84	0	421	227	0	9	0	18	0	2450	3
Portugal	63	1	441	207	2	29	3	17	0	5050	150
Puerto Rico	16	0	239	105	12	46	27	19	16	2493	235
Qatar	11	0	255	23	0	17	0	0	0	220	–
Réunion	2	0	43	18	4	3	3	0	0	546	165
Romania	84	0	368	247	0	25	0	19	0	3400	41
Russian Federation	269	22	703	628	13	58	0	41	0	21770	–
Rwanda	151	0	666	513	0	–	1	–	1	2288	26
Saint Helena	2	0	915	53	9	0	0	0	0	165	50
San Marino	13	0	137	–	0	9	0	3	0	–	–
Santa Lucia	9	0	169	50	4	17	5	2	0	1028	11

Country	Mammal species, total	Mammal species, endemic	Bird species, total	Bird species, breeding there	Bird species, endemic	Reptile species, total	Reptile species, endemic	Amphibian species, total	Amphibian species, endemic	Higher plant species, total	Higher plant species, endemic
São Tomé and Principe	8	4	111	63	25	16	7	9	8	895	134
Saudi Arabia	77	0	413	155	0	84	4	–	0	2028	–
Senegal	192	0	623	384	0	100	1	2	1	2086	26
Seychelles	6	2	170	38	11	37	14	12	11	250	182
Sierra Leone	147	0	622	466	1	–	1	–	2	2090	74
Singapore	85	1	141	118	0	140	0	24	0	2282	2
Slovakia	85	0	352	209	0	20	0	20	0	3124	92
Slovenia	75	0	360	207	0	25	0	20	0	3200	22
Solomon Islands	53	21	223	163	43	61	11	17	9	3172	30
Somalia	171	12	649	422	11	193	49	27	3	3028	500
South Africa	255	35	790	596	8	315	97	108	49	23420	–
Spain	82	4	506	278	5	53	11	28	4	5050	941
Sri Lanka	88	15	428	250	24	144	77	39	20	3314	890
St. Kitts and Nevis	7	0	99	32	0	10	0	1	0	659	1
St. Vincent and Grenada	8	1	129	108	2	16	4	3	0	1166	–
Sudan	267	11	937	680	1	–	8	–	1	3137	50
Surinam	180	2	673	603	0	151	0	95	8	5018	–
Swaziland	47	0	485	364	0	102	0	40	0	2715	4
Sweden	60	0	463	249	0	6	0	13	0	1750	1
Switzerland	75	0	400	193	0	14	0	18	0	3030	1
Syria	63	2	341	204	0	–	2	–	0	3000	–
Taiwan	63	11	445	160	14	80	14	36	14	3568	–
Tajikistan	84	1	348	–	0	44	0	2	1	–	–
Tanzania	316	15	1076	827	24	289	61	133	49	10008	1122
Thailand	265	7	915	616	2	298	37	112	21	11625	–
Togo	196	0	558	391	0	–	1	–	3	3085	–
Tokelau	0	0	15	5	0	7	0	0	0	26	–
Tonga	2	0	48	37	2	6	0	0	0	463	25
Trinidad and Tobago	100	1	433	260	1	70	3	26	3	2259	236
Tunisia	78	1	356	173	0	62	1	7	0	2196	–
Turkey	116	2	418	302	0	105	7	18	3	8650	2675
Turkmenistan	103	0	397	–	0	82	0	5	0	–	–
Turks and Caicos Islands	–	0	175	42	0	12	5	0	0	448	9
Tuvalu	–	0	27	9	0	–	0	0	0	–	–
Uganda	338	6	992	830	3	149	2	50	1	5406	–
Ukraine	108	1	400	263	0	21	1	17	0	5100	–
United Arab Emirates	25	0	360	67	0	37	1	–	0	–	–
United States of America	432	105	768	650	67	287	79	263	152	19473	4036
Uruguay	81	1	365	237	0	–	1	–	4	2278	40
Uzbekistan	97	0	427	–	0	64	0	2	0	4800	400
Vanuatu	11	2	111	76	9	20	4	0	0	870	150
Venezuela	323	19	1296	1340	40	283	66	245	122	21073	8000
Vietnam	213	9	761	535	10	187	46	80	27	10500	1260
Virgin Islands (American)	–	0	199	70	0	–	3	5	1	–	–
Virgin Islands (British)	3	0	199	70	0	18	3	5	1	–	–
Wallis Islands	1	0	23	25	0	–	0	0	0	475	7
Western Sahara	32	1	162	60	0	26	0	–	0	330	–
Western Samoa	3	0	60	40	8	8	0	–	0	737	–
Yemen	66	1	366	143	8	77	30	–	1	1650	135
Yugoslavia	96	0	380	224	0	70	0	21	0	4082	–
Zambia	233	3	731	605	2	145	3	65	1	4747	211
Zimbabwe	270	0	648	532	0	153	2	120	3	4440	95

Source XXI

0 = none
– = no data

White wine grapes

Works Cited and Used

Adams, J. / McShane, T.
The Myth of Wild Africa.
New York 1992

Bähr, J. et al.
Bevölkerungsgeographie.
New York 1992

Bailey, R. (ed.)
The true state of the planet.
New York 1995

**Baillie, J. / Groombridge, B. /
IUCN**
IUCN Red List of Threatened
Animals. Gland 1996

**Barthlott, W. /
Winiger, M. (ed.)**
Biodiversity. Berlin 1998

**Becker, U. / Ganter, S. /
Just, C. / Sauermost, R. (eds.)**
Lexikon der Biologie. Berlin 1994

**Begon, M. E. / Harper, J. L. /
Townsend, C. R.**
Ökologie. Heidelberg 1998

Berenbaum, M. R.
Blutsauger, Staatsgründer, Seiden-
fabrikanten. Heidelberg 1997

Bolton, M. (Hrsg.)
Conservation and the Use of
Wildlife Resources. London 1997

Bonner, R.
At the Hand of Man.
London 1993

Brand, S.
The Clock of the Long Now.
New York 1999

Breidenbach, J. / Zukrigl, I.
Tanz der Kulturen.
Munich 1998

**Bryant, D. / Burke, L. /
McManus, J. / Spalding, M. /
World Resources Institute**
Reefs at Risk. Washington 1998

Buchmann, S. / Nabhan, G.
The Forgotten Pollinators.
Washington 1996

Budiansky, S.
Nature's Keepers. New York 1995

Bundesamt für Naturschutz
Daten zur Natur. Münster 1996

Bundesamt für Naturschutz
Internationale Übereinkommen,
Programme und Organisationen
im Naturschutz. Bonn 1998

Bundesamt für Naturschutz
Materialien zur Situation der
biologischen Vielfalt in Deutsch-
land. Münster 1995

**Bundesministerium für
Umwelt, Naturschutz und
Reaktorsicherheit**
Konferenz der Vereinten
Nationen für Umwelt und
Entwicklung im Juni 1992 in
Rio de Janeiro. Dokumente
Klimakonvention, Konvention
über Biologische Vielfalt,
Rio Deklaration, Walderklärung.
Bonn (ohne Erscheinungsdatum)

**Bundesministerium für
Umwelt, Naturschutz und
Reaktorsicherheit**
Umweltpolitik. Konferenz der
Vereinten Nationen für Umwelt
und Entwicklung im Juni 1992
in Rio de Janeiro. Dokumente
Agenda 21. Bonn (ohne Erschein-
ungsdatum)

**Bundesministerium für
wirtschaftliche Zusammenar-
beit und Entwicklung (Hrsg.)**
Tropenwalderhaltung und
Entwicklungszusammenarbeit.
Bonn 1997

Calvin, H. W.
The River, That Flows Uphill.
New York 1986

Campel, N. A.
Biologie. Heidelberg 1997

Cohen, J. E.
How Many People Can the Earth
Support? New York 1996

Delort, R.
Der Elefant, die Biene und der
heilige Wolf. Munich 1987

**Deutsche Stiftung Welt-
bevölkerung / Gardner-Outlaw,
T. / Engelmann, R.**
Mensch, Wald. Stuttgart 1999

**Deutsche Stiftung
Weltbevölkerung**
Weltbevölkerung 1999.
Hanover 1999

Diamond, J.
Guns, Germs and Steel.
New York 1997

Dietl, W. / Meluhn, W.
Diemels Welt. Munich 1997

Dobson, A. P.
Conservation and Biodiversity.
New York 1996

Durrell, G.
Gerald Durrells Naturführer.
Munich 1983

Easterbrook, G.
A Moment on the Earth.
New York 1995

**Ellenberg, L. / Scholz, M. /
Beier, B.**
Ökotourismus. Heidelberg 1997

Ellis, R.
Mensch und Wal. Munich 1993

Eisenberg, E.
The Ecology of Eden.
New York 1998

**Falkenmark, M. / Widstrand, C. /
Population Reference Bureau**
Population Bulletin November
1992. Population and Water
Resources. Washington 1992

**Food and Agriculture Organi-
zation of the United Nations**
The State of the World
Fisheries and Aquaculture 1998.
Rome 1999

**Food and Agriculture Organi-
zation of the United Nations**
The State of the World's Forests
1999. Rome 1999

**Food and Agriculture Organi-
zation of the United Nations**
The Sixth World Food Survey.
Rome 1996

Freeman, J. D.
The Sun, the Genome and the
Internet. New York 1999

Freese, C.
Wild Species as Commodities.
Managing Markets and
Ecosystems for Sustainability.
Washington 1998

Gould, S. J.
Zufall Mensch. Munich 1991

**Green, M. J. B. / Murray, M. G. /
Bunting, G. C. / Paine, J. R. /
World Conservation
Monitoring Centre**
Priorities for Biodiversity
Conservation in the Tropics.
Cambridge 1997

**Groombridge, B. / World Con-
servation Monitoring Centre**
Global Biodiversity. London 1992

**Groombridge, B. /
Jenkins, M. D. / World Conser-
vation Monitoring Centre**
Biodiversity Data Sourcebook.
Cambridge 1994

**Groombridge, B. /
Jenkins, M. D. (ed.) /
World Conservation
Monitoring Centre**
Global Biodiversity – Earth's
living resources in the
21st century. Cambridge 2000

**Groombridge, B. /
Jenkins, M. D. / World Conser-
vation Monitoring Centre**
The Diversity of the Seas.
Cambridge 1996

Grzimek, B. (ed.)
Grzimeks Tierleben. Zurich 1971

Grzimek, B. (ed.)
Grzimeks Enzyklopädie.
Säugetiere. Munich 1988

**Guruswamy, L. / McNeely, J.
(ed.)**
Protection of Global Biodiversity.
Durham 1998

Hampicke, U.
Ökologische Ökonomie.
Opladen 1992

Harrison, P.
The Third Revolution.
London 1993

Hobusch, E.
Das große Halali. Berlin 1996

Hölldobler, B. / Wilson, E. O.
Ameisen. Basel 1995

Hölldobler, B. / Wilson, E. O.
The Ants. Berlin 1990

**Holdgate, M. / World Conser-
vation Union (IUCN)**
Can Wildlife Pay for Itself?
Gland 1993

**Institut der Deutschen
Wirtschaft**
IW-Umwelt Service May 1998.
Cologne 1998

Jacob, F.
Die Maus, die Fliege und der
Mensch. Berlin 1998

Jäger, H.
Einführung in die Umwelt-
geschichte. Darmstadt 1994

**Jelden, D. / Sprotte, I. /
Gruschwitz, M. / Bundesamt
für Naturschutz**
Nachhaltige Nutzung.
Münster 1998

Kegel, B.
Die Ameise als Tramp.
Zurich 1999

Keller, E. F.
Das Leben neu denken.
Munich 1998

Kümmer, K. et al.
Politische Ökologie, Special issue
10. Bodenlos. Munich 1997

Küster, H.
Geschichte der Landschaft in
Mitteleuropa. Munich 1995

Küster, H.
Geschichte des Waldes.
Munich 1995

Kurlansky, M.
Kabeljau. Munich 1997

Lange, D.
Untersuchungen zum
Heilpflanzenhandel in Deutsch-
land. Münster 1996

Leakey, R. / Lewin, R.
The Sixth Extinction.
New York 1995

Marggraf, R. / Streb, S.
Ökonomische Bewertung der
natürlichen Umwelt.
Heidelberg 1997

Margulis, L. / Sagan, D.
Leben. Heidelberg 1999

Maxeiner, D. / Miersch, M.
Öko-Optimismus. Dusseldorf
1996.

Maxeiner, D. / Miersch, M.
Lexikon der Öko-Irrtümer.
Frankfurt am Main 1998

**McNeely, J. /
Sochaczewski, P. S.**
Soul of the Tiger. Honolulu 1995

Mowat, F.
Der Untergang der Arche Noah.
Reinbek 1987

Müller, P.
Allgemeines Artensterben
– Ein Konstrukt?
Saarbrücken 1996

Nash, R.
Wilderness and the American
Mind. Yale 1982

Noin, D. / UNESCO
World Population Map. Paris 1997

North, R.
Life on a Modern Planet.
Manchester 1995

**Organization for Economic
Cooperation and
Development (OECD)**
OECD Environmental Data.
Compendium 1997. Paris 1997

**Organization for Economic
Cooperation and
Development (OECD)**
Towards Sustainable Develop-
ment. Environmental Indicators.
Paris 1998

Offenberger, M.
Von Nautilus zu Sapiens.
Munich 1999

Ostrom, E.
Governing the Commons.
Cambridge 1990

Pearce, D. / Moran, D.
The Economic Value of
Biodiversity. London 1994

Population Reference Bureau
World Population and the
Environment. Washington 1997

**Population Reference Bureau /
Cornelius, D. / Cover, J.**
Population & Environment
Dynamics. Washington 1997

**Population Reference Bureau /
Livernash, R. / Rodenburg, E.**
Population Bulletin. Population
Change, Resources, and the
Environment. Washington 1998

**Population Reference Bureau /
Gelbard, A. et al.**
Population Bulletin March 1999.
World Population Beyond Six
Billion. Washington 1999

Quammen, D.
The Song of the Dodo. New
York 1996

**Reaka-Kudla, M. / Wilson, D. E. /
Wilson, E. O. (Hrsg.)**
Biodiversity II. Understanding
and Protecting Our Biological
Resources. Washington 1996

**Redford, K. H. / Godshalk, R. /
Asher, K. / FAO**
What About the Wild Animals?
Rome 1995

Reichholf, J. H.
Das Rätsel der Menschwerdung.
Munich 1993

Reichholf, J. H.
Der Blaue Planet. Munich 1998

**Reid, W. V. / Laird, S. A. et al. /
World Resources Institute**
Biodiversity Prospecting
Using Genetic Resources for
Sustainable Development.
New York 1993

Reinicke, H.
Märchenwälder. Berlin 1990

Röser, B.
Grundlagen des Biotop- und
Artenschutzes. Landsberg on the
Lech 1990

Roth, H. / Merz, G. (Hrsg.)
Wildlife Resources. Berlin 1997

Schaefer, M. / Tischler, W.
Wörterbuch der Biologie.
Ökologie. Jena 1983

Schama, S.
Der Traum von der Wildnis.
Munich 1996

Scherzinger, W.
Naturschutz im Wald.
Stuttgart 1996

Simon, J. L.
The State of Humanity.
Oxford 1995

Sinclair, A. / Arcese, P.
Serengeti II. Dynamics.
Management and Conservation
of an Ecosystem. Chicago 1995

Silver, L. M.
Remaking Eden. New York 1997

**Spektrum der Wissenschaft,
Spezial**
Die dynamische Welt der
Ozeane. Heidelberg 1998

Statistisches Bundesamt
Bevölkerung und Wirtschaft
1872–1972. Stuttgart 1973

**Steiniger, F. / Türkay, M. /
Schminke, H. K. et al.**
Agenda Systematik 2000.
Frankfurt am Main 1996

**Stiftung Entwicklung
und Frieden**
Globale Trends 1998.
Frankfurt am Main 1997

**Swanson, T. M. /
Luxmoore, R. A. / World Con-
servation Monitoring Centre**
Industrial Reliance on
Biodiversity. Cambridge 1997

Ten Kate, K. / Laird, S. A.
The Commercial Use of
Biodiversity. London 1999

Tenner, E.
Die Tücken der Technik.
Frankfurt am Main 1997

Thomson, R.
The Wildlife Game.
Westville (South Africa) 1992

Treffi, J.
Lauter Gründe, warum die Welt
anders ist. Munich 1996

Tudge, C.
Letzte Zuflucht Zoo.
Heidelberg 1993

Tyler, I.
Geology & Landforms of the Kimberley. Como (Australia) 1996

Umweltbundesamt
Daten zur Umwelt. Berlin 1997

United Nations
1997 Demographic Yearbook.
New York 1999

**United Nations Development
Programme**
Bericht über die menschliche
Entwicklung 1998. Bonn 1998

Urania-Pflanzenreich.
Leipsic, Jena, Berlin 1995

Von Ungern-Sternberg, R.
Grundriß der Bevölkerungs-
wissenschaft (Demographie).
Stuttgart 1950

**Heywood, V. H. / Watson, R. T.
et al. / United Nations Environment Programme (Hrsg.)**
Global Biodiversity Assessment.
Cambridge 1995

**Whitmore, T. C. / Sayer, J. A.
(Hrsg.)**
Tropical Deforestation and
Species Extinction. London 1992

Wilson, E. O.
The Diversity of Life.
New York 1992

**World Business Council for
Sustainable Development /
World Conservation Union
(IUCN)**
Business and Biodiversity.
Geneva 1997

**World Business Council for
Sustainable Development**
Exploring Sustainable Develop-
ment. London 1997

**World Conservation
Monitoring Centre**
There is Only One Living Planet
and That's Why We Count.
Cambridge 1999

**World Conservation
Monitoring Centre / World
Wide Fund for Nature (WWF)**
Living Planet Report 1998.
Gland 1998

**World Conservation Union
(IUCN) / World Conservation
Monitoring Centre**
1997 United Nations List of
Protected Areas. Cambridge 1998

**World Conservation Union
(IUCN) / United Nations
Environment Programme (UN-
EP) / World Wide Fund
for Nature (WWF)**
Unsere Verantwortung für die
Erde. Gland 1991

World Health Organization
The World Health Report 1998.
Geneva 1998

World Health Organization
Health and Environment in
Sustainable Development.
Geneva 1997

**World Resources Institute /
World Conservation Union
(IUCN)**
Report of the Fifth Global
Biodiversity Forum. Gland 1997

**World Resources Institute /
The United Nations Environ-
ment Programme / The United
Nations Development
Programme / The World Bank**
World Resources 1998–1999.
A Guide to the Global
Environment. New York 1998

World Trade Organization.
Annual Report 1998. Geneva
1998

Worldwatch Institute
Zur Lage der Welt 1999.
Frankfurt am Main 1999

Wurm, S. A. / UNESCO
Atlas of the World Languages
in Danger of Disappearing.
Paris 1996

References for Illustrations and Sources for Plates

References for Illustrations

Scientific illustrations
Gundhild Eder

Pictograms
Fabian Nicolay

Maps
Mountain High Maps®
Copyright © 1993
Digital Wisdom Inc.

Deviations from these sources are indicated in the sources mentioned for each table.

Sources for Plates

Human Favorites
(pp. 18–19)
WCMC / Life-Counts-Survey /
Lone Pine Koala Sanctuary

Small Animals Shape the Earth
(pp. 20–21)
WCMC / M. R. Berenbaum:
Blutsauger, Staatsgründer, Seiden-
fabrikanten, 1997 / Tausch-Tremel
& Templ: Leben in der Antarktis,
1991 / U. Becker et al.,eds:
Lexikon der Biologie, 1994 /
Pro Futura u. WWF: Die Faszi-
nation der großen Zahl, 1998

The Ant as a Model of Success
(pp. 22–23)
B. Hölldobler and E. O. Wilson:
The Ants, 1990 (Illustrations
"Ant with Prey" and "Jaguar":
Christine Faltermayr)

A Changing World
(pp. 24–25)
A. P. Dobson: Conservation and
Biodiversity, 1996 / B. Grzimek,
ed.: Grzimeks Tierleben, 1971 /
R. Leakey and R. Lewin: The
Sixth Extinction, 1995 / F. R.
Paturi: Die Chronik der Erde,
1991 / Naturmuseum Sencken-
berg: Ausgerottete Vögel und
Säugetiere, 1983 (World map
"Continental Drift": Fabian
Nicolay)

Humans, a Career
(pp. 26–27)
W. H. Calvin: The River That
Flows Uphill, 1986 / J. E.
Cohen: How Many People Can
the Earth Support?, 1995 /
J. Diamond: Guns, Germs and
Steel, 1997 / J. H. Reichholf: Das
Rätsel der Menschwerdung, 1993
/ University of Marburg: World
Wide Web pages of the Computer
Science Department (Illustration
"Cranial Shapes": Fabian Nicolay)

Humans as Habitat
(pp. 28–29)
WCMC / U. Becker et al., eds:
Lexikon der Biologie, 1994 /
B. Grzimek, ed.: Grzimeks Tierleben, 1971 / W. B. Whitman et al.:
Prokaryotes – The Unseen Majority, Proceedings of the National Academy of Sciences (USA),
Juni 1998 / J. Graf: Tierbestim-
mungsbuch, 1961

Humans as Victims and Prey
(pp. 30–31)
WCMC / WHO / Florida Museum
of Natural History

Hot Spots of Biodiversity
(pp. 32–33)
WCMC / Deutsche Stiftung
Weltbevölkerung / Population
Reference Bureau / B. Grzimek,
ed.: Grzimeks Tierleben, 1971

Globalized Nature
(pp. 34–35)
WCMC / B. Kegel: Die Ameise
als Tramp, 1999 / New Scientist,
20 March 1999

Protected Areas:
Four Countries Compared
(pp. 36–37)
WCMC / L. Margulis u. D. Sagan:
Leben, 1997 / Umweltbundes-
amt: Daten zur Umwelt, 1997 /
Politische Ökologie, special issue
10/1995 / Drucksache des
Deutschen Bundestages 13-7400 /
Deutsche Stiftung Weltbevöl-
kerung

Survival in Zoos
(pp. 38–39)
IUDZG (The World Zoo Orga-
nization), IUCN: Die Welt-
Zoo-Naturschutzstrategie, 1997 /
W. Barthlott, Bundesamt für
Naturschutz: Botanische Gärten
und Biodiversität, 1999 /
A. P. Dobson: Biologische Vielfalt
und Naturschutz, 1997 / Botanic
Gardens Conservation Interna-
tional, 1996 / G. Nogge

**Animal Censuses in the
Serengeti**
(pp. 40–41)
WCMC / M. Borner
(Tanzania Wildlife Conservation
Monitoring)

**Bison Had to Give Way
to Cattle**
(pp. 144–145)
WCMC / National Bison
Association / National Cattlemen's Beef Association

**The World's Three Food
Sources**
(pp. 146–147)
WCMC / FAO / J. Diamond:
Guns, Germs and Steel, 1997

The Exploitation of the Seas
(pp. 148–149)
WCMC / FAO / WWF

Forests as an Economic Factor
(pp. 150–151)
WCMC / FAO / M. E. Begon et al:
Ökologie, 1998 (Illustration
"Oak Leaf": Fabian Nicolay,
world map of forested areas:
WCMC)

One Tree, Many Products
(pp. 152–153)
FAO / Philippine Coconut
Authority / Gesellschaft für
Technische Zusammenarbeit
(GTZ)

Genetically Altered Plants
(pp. 154–155)
T. M. Swanson, R. A. Luxmoore:
Industrial Reliance on Biodiver-
sity, 1997 / ISAAA (International
Service for the Acquisition
of Agri-Biotech Applications) /
J. Dorant et al.: Biotechnology
in the Public Sphere, 1998 /
R. Livernash and E. Rodenburg:
Population Change, Resources
and the Environment, Population
Bulletin Vol. 53/1

The Green Pharmacy
(pp. 156–157)
D. Lange: Europe's Medical
and Aromatic Plants. Their Use,
Trade and Conservation, 1998 /
M. J. Balick et al.: Economic
Botany, Bd. 49/2, 1995 /
Wissenschaftlicher Beirat der
Bundesregierung Globale
Umweltveränderungen: Welt
im Wandel, 1996 / CBD Report:
Access to Genetic Resources
and Benefit Sharing, 1995 /

K. ten Kate et al.: Benefit Sharing
Case Study, 1998 / K. ten Kate:
Biopiracy or Green Petroleum,
1995 / P. A. Cox et al.: Neue
Medikamente durch ethnobota-
nische Forschung, Spektrum der
Wissenschaft 8-1994

Nature as a Productive Force
(pp. 158–159)
WCMC / S. Buchmann,
G. Nabhan: The Forgotten Polli-
nators, 1996 / B. Conby: Délicieux
insectes, 1990 / A. P. Dobson:
Conservation and Biodiversity,
1996 / D. Jelden et al. (Bundesamt
für Naturschutz): Nachhaltige
Nutzung, 1998 / M. Holdgate
(IUCN): Can Wildlife Pay for
Itself?, 1993 / K. H. Redford et al.
(FAO): What About the Wild
Animals?, 1995 / J. H. Reichholf:
Der Blaue Planet, 1998 / H. Roth
u. G. Merz, Wildlife Resources,
1997 / T. M. Swanson und R. A.
Luxmoore (WCMC): Industrial
Reliance on Biodiversity, 1997 /
R. Thomson: The Wildlife
Game, 1992 / E. O. Wilson:
The Diversity of Life, 1992

Human's Housemates
(pp. 160–161)
K. ten Kate and S. A. Laird: The
Commercial Use of Biodiversity,
1999 / Fédération européenne
de l'industrie des aliments pour
animaux familiare (FEDIAF),
1996 / International Wolf Center,
1999

Life Counts Partners

WCMC
www.wcmc.org.uk

The World Conservation Monitoring Centre in Cambridge (UK) is the data-bank headquarters for international conservation. There, reports filed by field researchers all over the world are collected. This information constitutes the basis for certification regarding the protection and sustainable use of biological resources. Large global organizations, such as the UNEP, IUCN, and WWF, draw their data from the WCMC. WCMC experts were part of the Life Counts team. They also wrote the companion volume *Global Biodiversity – Earth's Living Resources in the 21st Century*, which is addressed to scientists (see following page).

IUCN
www.iucn.org

The World Conservation Union (IUCN is an abbreviation of this organization's earlier name) is the worldwide umbrella organization for conservation. Nine hundred governmental and non-governmental organizations from 138 countries are members of the IUCN. Founded in 1948, the IUCN became the largest global conservation institution. It seeks to help countries maintain their biodiversity and make use of it in an ecologically responsible way. The IUCN provided the leading idea for the world summit on sustainable development held in Rio de Janeiro, and has drafted many international agreements on conservation and sustainable use. The IUCN headquarters in Gland (Switzerland) also produces the global "red list" of endangered species.

UNEP
www.unep.org

The United Nations Environment Programme (UNEP) coordinates the United Nations' environmental activities. This organization, which has its headquarters in Nairobi (Kenya), collects and disseminates environmental information – for example, the "State of the Environment Reports" – and advises governments. It helps prepare international agreements and promotes cooperation between governments and the private business sector. Unlike the IUCN and the WCMC, the UNEP's area of activity includes not only biological resources (plants and animals) but also the whole realm of the environment (for instance, keeping the air clean and dealing with waste disposal problems). Like the other two organizations that participated in Life Counts, the UNEP is dedicated to the notion of "sustainable development."

Global Biodiversity – Earth's Living Resources in the 21st Century

In addition to Life Counts, a work addressed to specialists written by WCMC scientists will also be published in English.

On the occasion of the Rio de Janeiro conference in 1992, the WCMC brought out *Global Biodiversity*, a global inventory of biodiversity. After almost a decade of further research and analysis, the WCMC scientists are now publishing a new, comprehensive study, *Global Biodiversity – Earth's Living Resources in the 21st Century.*

This basic work, which is also comprehensible for lay people interested in the subject, is indispensable for research institutes, universities, governments, conservation organizations, and businesses that are concerned with biodiversity, conservation, ecosystems, or sustainable use of natural resources. It offers a complete scientific collection of data on the following subjects:
– The circulation of materials and energy flows in the biosphere;
– The diversity of organisms;
– Ecological and economic relationships between humans and nature;
– Ecosystems and their ranges of species;
– The global state of biodiversity at the beginning of the twenty-first century.

The text is accompanied by numerous tables and diagrams, as well as by specially-designed large-scale maps in color.

**Groombridge, B. /
Jenkins, M. D. ed. /
World Conservation
Monitoring Centre:**
Global Biodiversity
– Earth's living resources
in the 21st century.
Cambridge 2000
ISBN: 1 899628 150

First published in Germany in 2000 by Berlin Verlag, Berlin, Germany

Published simultaneously in Canada
Printed in England

FIRST EDITION

Library of Congress Cataloging-in-Publication Data

Life counts : cataloguing life on earth / Michael Gleich ... [et al.];
  translated by Steven Rendall.
    p. cm.
  ISBN 0-87113-846-8
  1. Animal populations—Statistics. 2. Animal populations—
Measurement. I. Gleich, Michael.

QL752 .L54 2002
591.7'88'0287—dc21   2001056613

Design by Fabian Nicolay

Atlantic Monthly Press
841 Broadway
New York, NY 10003

02 03 04 05   10 9 8 7 6 5 4 3 2 1